Studies in Computational Intelligence 475

Editor-in-Chief

Prof. Janusz Kacprzyk
Systems Research Institute
Polish Academy of Sciences
ul. Newelska 6
01-447 Warsaw
Poland
E-mail: kacprzyk@ibspan.waw.pl

T0140252

For further volumes:
http://www.springer.com/series/7092

Gemma C. Garriga

Formal Methods for Mining Structured Objects

 Springer

Gemma C. Garriga
INRIA Lille Nord Europe
Parc Scientifique de la Haute Borne
Villeneuve d'Ascq
France

ISSN 1860-949X   ISSN 1860-9503   (electronic)
ISBN 978-3-642-43063-3   ISBN 978-3-642-36681-9 (eBook)
DOI 10.1007/978-3-642-36681-9
Springer Heidelberg New York Dordrecht London

Printed on acid-free paper

Springer is part of Springer Science+Business Media (www.springer.com)

Preface

In the field of knowledge discovery, graphs of concepts are an expressive and versatile modeling technique providing ways to reason about information implicit in a set of data. Interesting examples of this can be found under the classical mathematical theory of formal concept analysis, dedicated to the construction of a lattice of concepts by defining a Galois connection on a binary relationship. Typically, nodes of this graph represent patterns, such as sets of items, and edges represent the relationships of specificity among them. In this manuscript we will study such graph of concepts under the more complex case of data that comes in a set of structured objects; e.g. a set of sequences, trees or even of graphs. As a natural step towards the general characterization, we first focus on the mining of sequential data and, for this case, we will study the formalization of a lattice of closed sets of sequences.

We will motivate that this lattice is an interesting combinatorial object from which to derive justified methods for current sequential mining problems. The first set of results from the lattice focuses on the characterization of logical implications with order. We propose a notion of association rules and prove that they can be formally justified by a purely logical characterization, namely, a natural notion of empirical Horn approximation for ordered data, which involves background Horn conditions; these ensure the consistency of the propositional theory obtained with the ordered context. We also discuss a general method to calculate these rules that can be easily incorporated into any algorithm of discovery of closed sequential patterns. The second set of results corresponds to the identification of partial order structures from the input sequences. We will show that the maximal paths of the closure of such partial orders can be derived from the closed sets of sequences of our lattice. This theoretical result allows for the construction of partial orders by gluing the proper closed sequential patterns into higher-level structures. This is formally justified through a graph transformation from our lattice of closed sets of sequences into an isomorphic lattice of closed partial orders. The main proof appeals to basic operations of category theory.

This is a research monograph book containing mostly published results but also many unpublished results. We hope that the proposed analysis will provide new insights and understanding into the mining of, not only sequences, but also other structured data without cycles. The results presented in this book will be coupled with examples and empirical experiments; these do not intend to be exhaustive but want to illustrate that the outlined theoretical contributions produce appealing results also in practice.

January, 2013   Gemma C. Garriga

Acknowledgements

This manuscript is mostly the outcome of the research performed during my PhD thesis. I am therefore very grateful to my supervisor and main collaborator during that time, José Balcázar; his contributions, insights and discussions are also part of the following pages. Special thanks also to my family and friends.

Contents

Chapter 1
Introduction

The idea of extracting knowledge from sets of data emerged back in the 90s, motivated by the decision support problem faced by several retail organizations that due to several technological advances, were able to store massive amounts of sales data. At that point, the research field of knowledge discovery started as an active area of investigation, and data mining, one of the most challenging steps of the process, was meant to provide efficient algorithms and techniques to automatize the exploratory analysis of the data. In its simplest form, such data is viewed as a set of transactions where each transaction is a set of items (attributes of the database), that is, a simple binary relation. This representation is known popularly as market-basket data.

One natural way of representing knowledge in this context is to look for causal relationships, where the presence of some facts suggests that other facts follow from them. One of the reasons for the success of the association rule framework is that in the presence of a community that tends to buy, say, sodas together with the less expensive spirits, a number of natural ideas to try to influence the behaviour of the buyers and profit from the patterns, easily come up. Approaches to find such associations started long before the area of knowledge discovery became so popular. For example, Duquenne and Guigues in [46] and also Luxenburger in [93], studied bases of minimal nonredundant sets of rules from which all other rules can be derived. The former studied these bases for association rules with 100% confidence, and the latter association rules with less than 100% confidence, but neither of them considered the support of the rules, i.e. the number of transactions in the data supporting the rule. Nowadays, this task is widely known as the association rule mining problem, and became very popular since it was reformulated by Agrawal et al. in [3]. The reformulation made by Agrawal et al. introduced this notion of support, allowing for the pruning of those rules whose number of occurrences in the data was not over a user-specified threshold.

Taking this association rule mining problem, there is a rich variety of algorithmic proposals whose strategy is to look for the frequent itemsets in the data, i.e., those sets of items with numbers of occurrences over a threshold, and then, constructing implications between these discovered frequent itemsets. The most well-known of these algorithms is Apriori [4]; it traverses the search space in a breadth-first fashion,

G.C. Garriga: *Formal Methods for Mining Structured Objects*, SCI 475, pp. 1–11.
DOI: 10.1007/978-3-642-36681-9_1   © Springer-Verlag Berlin Heidelberg 2013

using the antimonotonicity property of the support to prune unnecessary candidates. After this first proposal, many other algorithmic strategies and methods emerged to improve the efficiency of Apriori: e.g. some of them suggested new structures to compact the original database into main memory, others proposed a way to traverse the search space in a best-first fashion, or performing the mining over a sample of the original transactions instead of all the data, or even some publications worked with parallel algorithms. Among many others, the following works are relevant [2, 20, 9, 10, 27, 28, 64, 69, 74, 72, 77, 91, 90, 101, 112, 123, 142, 145, 138, 147]. For a recent survey on the algorithmic trends of this problem see [71]. To complement these algorithmic advances with theory, in [66, 67] the connection between association rules and hypergraph transversals was presented; the authors also gave complexity bounds to discover those maximal sets with their hypergraph formulation.

Soon after the publication of the aforementioned algorithms, the following problem was how to reduce the huge number of association rules that were extracted by the algorithms. Different criteria were needed to make a judgment whether the extracted implication contained useful information. A classical way to rank the final rules is by means of statistical metrics (e.g. confidence [3, 93], conviction [27], lift [26] and so on). There are a large number of proposals as to how to measure the strength of implication of a rule, yet criticisms of various forms can be put forward for any measures; e.g. one of the criticisms for lift is its symmetry, which makes it impossible to orient rules. Surveys, with appropriate references, are given in [21, 22, 55, 48, 76, 115, 119, 127]. Recent advances on defining the significance of itemsets and rules are, among many others, e.g. [49, 121, 94, 130].

A complementary approach to ranking rules with statistical metrics consists of generating a basis of association rules from which the rest can be derived. As mentioned before, this approach was initially studied by Duquenne and Guigues in [46], and later by Luxenburger in [93]; yet, they did not consider any notion of minimum support for the rules. This idea evolved towards considering only those frequent *closed itemsets* instead of all the frequent itemsets when first mining the data, and after that, generate only those rules indicated by the closure system (see e.g. [18, 39, 102, 120, 126, 133, 144, 137, 141]). Using a similar idea, the work in [40] introduced a new rule of inference and defined the notion of association rule cover as a minimal set of rules that are non-redundant with respect to this new rule of inference. Other complementary ideas are the non-derivable itemsets of [31, 32], relying on a complete set of deduction rules. More recent works are [12, 11, 82]. Even if these approaches of covering rules or compressing patterns also result effective in practice, we will focus here on the theory related to closed patterns.

As we shall see, closed itemsets are particularly interesting from a theoretical point of view due to their mathematical foundations based on formal concept analysis and concept lattices. Broadly speaking, this theory is based on the definition of a Galois connection for a binary relation between a set of objects and a set of items, that is, the original data. This Galois connection enables a closure system, i.e. a complete lattice (a Hasse diagram) of formal concepts. Each one of these concepts captures the information of closed itemsets, hence implications, in the data.

The theory of Galois lattices has proved to be an expressive technique to reason about the binary data.

Nonetheless, for many real applications data are represented in more complex structures, such as sequences, trees or graphs. Squeezing these structures into a single relation may lead to a loss of information, and so, we require specific techniques and formalizations different from the ones commonly applied to single normalized tables. The most basic type that data can exhibit corresponds to the sequential categorical domain, i.e. elements follow in a specific sequential order. These elements in the sequence may have a simple form, such as a single item, or also have a more complex structure, such as sets of items or even a hierarchical organization. This is a complex task due to the combinatorial explosion of searching and generating new patterns, which may range from a plain structure (sequential subsequences) to a more complex tree-like form (such as partial orders).

The sequential mining problem was initially posed by Agrawal and Srikant in [5], and most of the work has focused on providing efficient algorithms for mining frequent patterns of various forms in the sequential data; e.g. works such as [96, 95] are dedicated to the mining of partial orders, and others such as [73, 104, 117, 139] to the mining of frequent subsequences. Recent advances on mining algorithms and probabilistic models for sequential data are surveyed in [45].

In this manuscript, we consider that mining a set of sequences is the first natural step to work towards the closure-based analysis of complex structured objects; the goal here is to provide a theoretical insight into the formalization this domain via formal concept analysis and lattice theory. Intuitions obtained in the sequential case will give a good intuition into other complex combinatorial mining problems, such as having a set of graphs as our input data. This first chapter aims at giving an overview to the specific tasks of mining of sequences that we will be considering through the rest of the book.

1.1 Analysis of Sequences

Let $\mathscr{I} = \{i_1, \ldots, i_m\}$ be a fixed set of items. A subset $I \subseteq \mathscr{I}$ is called an itemset. Formally, we deal with sequential categorical data, described as a collection of ordered transactions $\mathscr{D} = \{d_1, d_2, \ldots, d_n\}$, where each d_i is a *sequence* of finite length. Sequences $d_i \in \mathscr{D}$ will be called input sequences or transactions in this documentation.

We consider a *sequence* to be an ordered list of itemsets. It can be represented as $\langle (I_1)(I_2)\ldots(I_n) \rangle$, where each I_i is a subset of $\mathscr{I}$, and I_i comes before I_j if $i \leq j$. Note that we model each element of the sequence, not as an item, but as an itemset. Without loss of generality we assume that the items in each itemset are sorted in a certain order (such as alphabetic order); and to simplify, itemsets will be displayed without the curly brackets, i.e. ACD represents $\{A, C, D\}$. The universe of all the possible sequences will be denoted by $\mathscr{S}$. Our set of input sequences is a subset of this universe, i.e. $\mathscr{D} \subseteq \mathscr{S}$.

Fig. 1.1 Example of a sequential database $\mathscr{D}$

Seq id	Input sequences
d_1	$\langle (AE)(C)(D)(A) \rangle$
d_2	$\langle (D)(ABE)(F)(BCD) \rangle$
d_3	$\langle (D)(A)(B)(F) \rangle$

This description of $\mathscr{D}$ corresponds exactly to the model of sequences originally proposed by Agrawal and Srikant in [5], and subsequently followed by other works on mining sequential data. This fits exactly in the context of having a *sequential database*, such as a dataset of customer shopping sequences, but it can also fit in case of dealing with *time-series data*, such as alarms in a telecommunication network. In this latter case, the long string of events can be divided into several sliding windows representing each one a piece of $\mathscr{D}$, so that all the transactions would have the same fixed length. A small synthetic example of $\mathscr{D}$ is presented in Figure 1.1.

Formally, we will need some basic operations on sequences.

Definition 1.1 ([5, 117]). We say that a sequence $s = \langle (I_1) \dots (I_n) \rangle$ is a **subsequence** of $s' = \langle (I_1') \dots (I_m') \rangle$, i.e. $s \subseteq s'$, if there exist integers $1 \le j_1 < j_2 \dots < j_n \le m$ s.t. $I_1 \subseteq I_{j_1}', \dots, I_n \subseteq I_{j_n}'$; then, we also say that s is **contained** in s'.

For example, the sequence $\langle (C)(D) \rangle$ is contained in $\langle (AC)(D)(B) \rangle$, but it is not contained in $\langle (CD)(A) \rangle$. Reciprocally, we also define that $s \subset s'$ when $s \subseteq s'$ and $s \ne s'$. A sequence is *maximal* in a set of sequences if it is not contained in any other sequence of the set.

Notice that the notion of subsequence that we just introduced is taken directly from the initial work of Agrawal and Srikant in [5, 117], and this has become the commonly accepted formalization. However, we may think of another interpretation of this operation, which is considering equality in the indexes $1 \le j_1 \le j_2 \dots \le j_n \le m$, thus allowing a sequence to be totally included in one of the itemsets of another sequence. Under this new interpretation a sequence such as $\langle (A)(D)(B) \rangle$ would be included in another such as $\langle (ACD)(B) \rangle$. Formally, this may lead to the disadvantage of letting sequences of infinite length be included into finite ones. For example, an infinite sequence of A's such as e.g. $\langle \dots (A)(A) \dots \rangle$, would be allowed to be included in $\langle (A) \rangle$. This seems an unnatural choice for our applications, where the database $\mathscr{D}$ is composed only of finite input sequences by definition. Here we remain faithful to the interpretation given by Agrawal and Srikant, also to ease the translation of our results w.r.t. former works. We will analyze below the consequences of this choice.

The transaction identifier list of a sequence s w.r.t. $\mathscr{D}$, denoted $tid(s)$, is the list of input sequence identifiers from $\mathscr{D}$ where s is contained, e.g. $tid(\langle (AE)(D) \rangle) = \{d_1, d_2\}$ for the data in Figure 1.1. For short, we will write identifiers with natural numbers, that is, $tid(\langle (AE)(D) \rangle) = \{1, 2\}$; later, this simplification of the notation will allow for a comparison with the identifiers used in formal concept analysis. The *support* of a sequence s, denoted as $supp(s)$, is the number of occurrences of s in $\mathscr{D}$; e.g. $supp(\langle (AE)(D) \rangle) = |tid(\langle (AE)(D) \rangle)| = 2$.

Associated to the analysis of sequences there are different tasks and problems. Among all the tasks that one could imagine on such data, we will give a brief overview of the following ones: mining closed sequential patterns, summarizing the data by means of partial orders, mining association rules with order and clustering input sequences.

1.1.1 Mining Closed Sequential Patterns

A relevant task of the sequential mining problem is the identification of frequently-arising patterns or subsequences; in other words, those subsequences in $\mathscr{D}$ whose support is over a user-specified value. These frequent sequential patterns turn out to be useful in many domains, for instance in the anomaly detection for computer security ([85, 88, 89]). Managing sequential patterns and counting their support in $\mathscr{D}$ is a challenging task since one needs to examine a combinatorially explosive number of possible frequent patterns. Many studies have contributed with algorithms for this problem, e.g. [52, 97, 104, 73, 117, 139]. See [45] for a thorough list of references. Unfortunately, there are important cases where the number of frequent patterns is too large for a thorough examination and the algorithms face several computational problems; these include the cases of considering a very low threshold or a dense database (i.e. with high correlation between the items of the input sequences).

Proper solutions to this were initially proposed by Yan, Han and Afshar in [135]. They consists of mining just a compact and more significant set of sequential patterns called the *closed sequential patterns*. This idea parallels the notion of closed itemsets in a binary database, and indeed, both are defined as patterns not extendable to others with the same support. Formally:

Definition 1.2. Given a database $\mathscr{D}$, a sequence $s \in \mathscr{S}$ is **closed** (also known as a **closed sequential pattern**) if there exists no sequence s' with $s \subset s'$ s.t. $supp(s) = supp(s')$.

For instance, taking data from Figure 1.1, we have that $\langle (A)(F) \rangle$ is not closed since it can be extended to $\langle (D)(A)(F) \rangle$ in all the input sequences where it is contained. However, sequences such as $\langle (D)(A) \rangle$ or $\langle (AE)(C) \rangle$ are closed since they are maximal among those others with the same tid list. The set of all the closed sequences and their tid lists from data in Figure 1.1 are presented in Figure 1.2.

For the sake of comparison, in Figure 1.3 we also provide the list of closed sequences that would be derived from the same data in Figure 1.1, yet considering the redefined notion of subsequence that we suggested above. Note that by accepting the total inclusion of one sequence into one itemset we get a more compacted set of final patterns: the classical interpretation of Agrawal and Srikant leads to 9 closed sequences, whereas with the new interpretation we get 6 closed patterns. Despite the potential of this redefinition in practice (under the proper formalizations so as to avoid an infinite number of closed patterns in the set of data), the present document is fully dedicated to the classical interpretation of subsequence.

Fig. 1.2 All closed sequences derived from the data in Figure 1.1

Tid list	Closed Sequential Patterns
{1}	$\langle (AE)(C)(D)(A)\rangle$
{2}	$\langle (D)(ABE)(F)(BCD)\rangle$
{3}	$\langle (D)(A)(B)(F)\rangle$
{1,2}	$\langle (AE)(C)\rangle$
{1,2}	$\langle (AE)(D)\rangle$
{2,3}	$\langle (D)(A)(B)\rangle$
{2,3}	$\langle (D)(A)(F)\rangle$
{2,3}	$\langle (D)(B)(F)\rangle$
{1,2,3}	$\langle (D)(A)\rangle$

Fig. 1.3 All closed sequences derived from data in Figure 1.1 with a new interpretation of the subsequence operation, as explained above

Tid list	Closed Sequential Patterns
{1}	$\langle (AE)(C)(D)(A)\rangle$
{2}	$\langle (D)(ABE)(F)(BCD)\rangle$
{3}	$\langle (D)(A)(B)(F)\rangle$
{1,2}	$\langle (AE)(C)(D)\rangle$
{2,3}	$\langle (D)(A)(B)(F)\rangle$
{1,2,3}	$\langle (D)(A)\rangle$

In [135], Yan, Han and Afshar present the first of a series of algorithms for mining closed sequences in $\mathscr{D}$ over a minimum support, named CloSpan. Later other algorithms followed up to improve its efficiency (e.g. TSP [125] or BIDE [131, 132] or also [38, 109] among many others). The way these algorithms work to identify the closed sequences in the data and their frequency is actually irrelevant for our purposes, and, in general we will use CloSpan as the representative of this group of algorithms that mine closed sequences. Mainly, we consider that the interest in using closures relies on their theoretical characterization: while closed itemsets set up their basis on classical formal concept analysis, there is no such direct formal characterization of the ordered counterpart.

1.1.2 Mining Partial Orders

Alternatively to the mining of frequent sequential patterns, other approaches were designed to go beyond the plain structure of sequential patterns and consider tree-like patterns to summarize the input sequences. The importance of mining more complex structures from sequential data was first argued by Mannila et al. in [96]. The authors consider the mining of frequent episodes, i.e. collections of events occurring frequently together in the input sequences. Episodes are formalized as

Fig. 1.4 Example of a partial order (also called "hybrid episode")

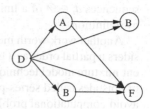

acyclic directed graphs, and they can be classified into serial episodes (total orders), parallel episodes (trivial orders), and finally, hybrid episodes (general partial orders). In Figure 1.4 there is an example of a general partial order compatible with the second and third input sequences of data in Figure 1.1 (compatibility will be defined formally in chapter 5). Moreover, notice that this partial order is the most specific one for these two input sequences: intuitively, no new node or new edge between the existing nodes can be added to the structure to make it more informative; that is, no other partial order can summarize better those two input sequences.

From the algorithmic perspective, the work in [96] discusses two different approaches for the discovery of frequent episodes in $\mathscr{D}$: Winepi and Minepi. The approach called Winepi is intended to look for frequent episodes following the Apriori scheme, that is, by sliding a window of fixed width along the event sequence. A complete pass along the data is used to compute the support of current episode candidates and, after each pass, new larger episodes are generated as long as the antimonotonicity property of support keeps them active. So, at the end of the process Winepi has discovered all the frequent episodes fitting in the window.

The problem arises with the complexity of managing these structures and the combinatorial explosion to tackle all the cases. Winepi performs two complex operations: first, generating a new set of candidates out of smaller episodes, and second, identifying the compatibility of episodes in each transaction to update the support. In case of mining dense data, and specially, when dealing with hybrid episodes or with noninjective episodes (those where the labels of the nodes can be repeated), the algorithm incurs in a substantial runtime overhead. Apart from this algorithmic overhead, the number of the final discovered episodes is quite large and many of them could be considered redundant: e.g. many of the final parallel episodes may be less informative than some of the serial episodes, and in turn, many serial episodes may be less informative than the hybrid episodes. Yet another objection, if the chosen window is not wide enough, the final discovered episodes will be simply overlapped parts of a longer, more informative episode that cannot fit in the window.

Alternatively to Winepi, the idea of mining unbounded episodes was proposed as a way to solve this latter problem. So, in [96] the same authors present Minepi, which follows again an strategy similar in spirit to Apriori. Another similar attempt presents episodes with gap constraints [53, 98, 79]; or a different idea of frequency that avoids overlapping episodes is presented in [86]. However, these approaches do not fit directly in the transactional model presented here: we are considering input sequences $d_i \in \mathscr{D}$ of a finite length, and unavoidably, our patterns will be bounded by definition.

Another work worth mentioning is [95] by Mannila and Meek. This method considers a partial order as a generative model for a set of sequences and applies different mixture model techniques. However, they must restrict the attention to a subset of episodes called series-parallel partial orders (such as series-parallel digraphs) to avoid computational problems. Other interesting papers working with episodes are [7] or [41, 124], they deal with serial episodes more than hybrid structures. In general, identifying such hybrid structures directly from the data is a complex task due to the combinatorial nature of the problem.

Other type of probabilistic approaches to identify episode-like models, such as hidden Markov models, are succinctly described in [45]. Finally, the notion of closed episodes has been also studied by several authors, thus proposing efficient algorithms to extend the notion of closed sequential patterns defined above [146, 122, 106, 105]. More formal notions of a closed episode (closed partial order) will be given in chapter 5; our interest is rather to theoretically characterize these notions which are widely used in practice.

1.1.3 Mining Association Rules with Order

A first antecedent to the mining of association rules in sequences is again the work of Mannila et al. in [96]. Actually, we can see serial episodes (total orders) as sequences, and then, an association rule with order has the form $s \rightarrow s'$ with $s \subseteq s'$. In other words, we have a sequence implying another sequence, where the antecedent must be contained in the consequent. Of course this idea of implication is based in the classical propositional framework of itemsets, and thus, it can be extended to any hybrid structure, not only sequences.

As it happens with the unordered case, the number of constructed rules can be quite large and difficult to examine. Again it is possible to compute an interestingness measure over the rule with order, such as confidence in [96]. Also other deterministic solutions are naturally possible, e.g. the work in [75] proposes to study just those representative rules obtained through implications of serial closed episodes.

It turns out that serial closed episodes of [75] (which are total orders by definition) are exactly the closed sequences defined in [135] and mined by CloSpan. There is an important algorithmic difference though: the algorithm presented in [75] follows an Apriori scheme, while CloSpan follows the tree structure of PrefixSpan [104]; alternative algorithms have appeared in [125] (TSP) or [131] (BIDE), or recently [38].

In chapter 4 we will consider a generalization of all these association rules by presenting a novel notion of implications where a set of sequences (total orders) in the antecedent imply a single sequence in the consequent. In our case, these sequences in the antecedent may not be necessarily a subsequence of the sequence in the consequent, turning into more informative the predictive rule. Moreover, one can prove that the set of all these rules exhibits an interesting logical characterization derived from the closure system of sequences.

1.1.4 Clustering Input Sequences

Clustering is the task of grouping together objects into meaningful subclasses. Here transactions in $\mathscr{D}$ can be considered a set of objects described by sequential attributes. The goal is to group objects in $\mathscr{D}$ into different clusters, by using specific discriminating features. This can be useful in different contexts, for example when considering order in the shopping bags of the market basket data, which leads to groups of customers with similar purchasing patterns. One of the key steps in clustering algorithms is the method for computing the similarity between the objects being clustered. The work in [68] by Guralnik and Karypis, considers as discriminating features all the sequential patterns over a certain support with length between two values given by the user. The critical step is then to project the new incoming sequences into the new feature space, so that they must restrict the space to only a reduced set of features. Other notions of clusters for sequences, or representatives of sets of sequences, are based on hidden Markov models, e.g. [116, 87]. In the next chapters we will show that the Galois lattice of closed sequences represents naturally a hierarchical organization of the final clusters.

1.2 Overview of this Book

The main goal of the book is to study the formalization of closed structured patterns in the sequential data. Plain frequent closed sequential patterns proved to be useful in many ways: first, the user needs to examine fewer patterns obtained as an output of the mining algorithms; second, hitting with the right minimum support threshold is not so important, for example, mining all the subsequences in $\mathscr{D}$ with a threshold close to zero is unrealistic and it does not provide useful information, but the set of all closed sequences is not so dramatic and still gives an overall idea of the whole database. In general, our main motivation is that closure systems define a reduced search space with the potential to be characterized by a sound mathematical background based on formal concept analysis. We will completely characterize this closure space of structured patterns for the sequential data, while working at the same time towards the analysis of other structured objects.

First, it is interesting to realize that the set of closed sequential patterns does not represent all the particularities hidden in the sequential data. Formally, there can exist two closed sequences s and s' such that they occur in the same transactions, so that $tid(s) = tid(s')$, but $s \nsubseteq s'$ and $s' \nsubseteq s$. In other words, contrary to the case of closed itemsets in binary data, here there is no unique closure representing a given set of ordered transactions. By way of example, the closed sequences $\langle (AE)(C) \rangle$

and $\langle (AE)(D) \rangle$ from Figure 1.2 occur in the same set of transactions, and none of them can be considered "better" than the other, they simply coexist together.

Indeed, the following research project started with the aim of studying all these particularities of sequential data, by using formal concept analysis as a formalization tool. We rely on the fact that formal concept analysis is a methodology of data analysis and knowledge representation with the potential to be applied to a variety of fields. Indeed, for the unordered context of binary transactions this theory has given already very interesting results, see e.g. [18, 39, 102, 120, 126, 133, 144, 137, 141]. Here, we show that this framework can be very powerful to unify different sequential mining tasks and provide, not only theoretical formulations that help in the understanding, but also new ideas to work towards efficient algorithmic solutions.

As a consequence of the foundations provided for the sequential case, we will show how these contributions can be extended to other kind of structured data which do not contain cycles, mainly represented as partial orders or trees.

Organization

As the simplest case towards the analysis of different structured objects, we take a set of input sequences and we develop a formal framework based on a closure system generated from the foundations of formal concept analysis. This requires to define the proper Galois connection adapted to sequences, and from here, it is direct to construct the concept lattice capturing the particularities and relationships of our data. Then, we use this combinatorial object to formally propose justified methods for the tasks described in this introductory chapter. We organize this document as follows.

Chapter 2 introduces some preliminaries on formal concept analysis for binary data. These foundations will be the first step to understand the characterization of our closure system for sequences in the following chapters.

Chapter 3 sets the mathematical formulation defining our closure system on sequences. This will be the model used as the basis of the subsequent contributions and it corresponds to a lattice of closed sets of sequences. In this chapter we will also prove some basic results that characterize the construction of such a lattice, and the relationship with the classical closed sequential pattern mining. We will provide simple algorithmic schema for the construction of the lattice.

Chapter 4 deals with the problem of defining association rules in ordered data. We will present a novel notion of implication that can be derived from the closure system: a set of sequences imply an individual sequence in the data, and each one of the sequences in the antecedent is not necessarily contained in the sequence of the consequent. We prove that these rules can be formally justified by a purely logical characterization, namely a natural notion of empirical Horn approximation for ordered data which involves specific background Horn conditions. We resolve a way to calculate these implications with order by means of generators of each closed set of sequences in the lattice. The proposed method can be incorporated to current algorithms for mining closed sequential patterns, of which there are already some in the literature.

Chapter 5 and 6 address the task of summarizing the input sequences by means of partial orders. As a main result we show that the maximal paths of the closed partial orders in the data can also be derived from the nodes of our lattice model. This leads to an interesting algorithmic simplification: we are avoiding the complexity of the mining operation of these structures directly from the input transactions; now we can obtain the hybrid partial orders by just gluing conveniently their maximal paths, which correspond to closed sequential patterns. This result gives the possibility of developing both theoretically and algorithmically the identification of closed partial orders. The work presented in this chapters focuses on the theoretical part of this result, by showing that the identification of a set of maximal paths into a hybrid structure can be characterized with basic operations of category theory [1], through coproducts and colimits. This is not an easy task, specially in the case of considering repeated items in the input sequences. To ease the understanding of the contributions we develop this contribution in two steps. In chapter 5 we simplify the problem to the case of dealing with partial orders where labels cannot be repeated, and in chapter 6 we address the general case of allowing for repeated labels. In chapter 6 we leave as an open question one interesting extension of a result in chapter 5 regarding the property of maximal specificity of our partial order. Assuming this result as a working hypothesis, we can prove the necessary properties that ensure the isomorphy between the closure system of sequences and the corresponding closure system of partial orders.

Chapter 7 develops the extension of the results obtained for the sequential case to other complex structured objects without cycles, which can be very well represented as partial orders. Broadly speaking, each input partial order will be transformed here into a set of sequences corresponding to its maximal paths. Then, a proper notion of subpattern in the new transformed data will allow to directly generate closed structures on the basis of the former theoretical contributions for sequences. Also, this chapter provides the necessary experimental evaluation required to justify our contributions from the more practical point of view.

Chapter 8 concludes and provides a brief overview of the contributions. Most of the results from this book can be found in the following publications [56, 59, 58, 54, 17, 16, 15, 57].

Chapter 2
Preliminaries

Formal concept analysis, briefly written as FCA from now on, is based on the mathematical theory of complete lattices (see [51] as a main reference, and also [42] for lattice theory). This theory provides an elegant mathematical framework that has been used in a large variety of fields of computer science, such as in knowledge discovery, where the obtained graph of concepts has proved to be a powerful formalization tool. Moreover, FCA has eased the study of dependencies in many-valued contexts [8, 36, 43], or as mentioned in the introduction, association rules in a binary database, e.g. see [39, 120, 144, 137, 141], among many other applications. In this chapter, we present a brief introduction and a comprehensive view of the methods derived for data mining within this theory. For a more thorough introduction from the computer science perspective see [35].

2.1 Contexts, Concepts and the Concept Lattice

Let $\mathscr{I}$ be the fixed set of items, as defined in the introductory chapter. For the main case of interest in knowledge discovery, a *formal context* consist of a triple $(R, \mathscr{O}, \mathscr{I})$, where R is a binary relation between the set of objects and the set of items, that is, $R \subseteq \mathscr{O} \times \mathscr{I}$. We may think of a context as a set of bit vectors where each bit on corresponds to an object having that particular item. An example of such a relation is shown in Figure 2.1(b). Indeed, the popular market transactional data can be treated directly as a formal context, as it follows from Figure 2.1: each transaction (also called tuple) t_i in the data is equivalent to the object o_i.

The first step towards the lattice construction is to define a Galois connection between two *derivation operators*: one mapping a set of objects into a set of items and the other mapping a set of items into a set of objects. To simplify notation we denote objects in $\mathscr{O}$ as the natural numbers from 1 to n (where n is the size of the database).

G.C. Garriga: *Formal Methods for Mining Structured Objects*, SCI 475, pp. 13–19.
DOI: 10.1007/978-3-642-36681-9_2   © Springer-Verlag Berlin Heidelberg 2013

Fig. 2.1 Equivalence of the
transactional data with a
binary context

Trans. Id	Sets of items
t_1	ACD
t_2	ABC
t_3	AB

(a) Collection of stan-
dard transactional data

Objects	A	B	C	D
o_1	1	0	1	1
o_2	1	1	1	0
o_3	1	1	0	0

(b) Formal context
in FCA theory

- For a set of objects $O \subseteq \mathcal{O}$, it is common to define:

$$\alpha(O) = \{i \in \mathcal{I} \mid \text{item } i \text{ is contained in } t_o, \text{ for all } o \in O\}$$

  E.g. $\alpha(\{1,2\}) = AC$ for the context in Figure 2.1. Operator α can be regarded
  as the intersection of the set of objects, corresponding each one of them with an
  itemset; for example, $\alpha(\{1,2\})$ coincides with the intersection of two itemsets
  ACD and ABC, i.e. to transactions t_1 and t_2 respectively. The output of operator
  α will be always one single itemset.
- Dually, for a set of items $I \subseteq \mathcal{I}$, it is common to define:

$$\beta(I) = \{o \in \mathcal{O} \mid t_o \text{ contains the item } i, \text{ for all } i \in I\}$$

  For example, $\beta(A) = \{1,2,3\}$. In other words, this function can be also under-
  stood as the transaction identifier (tid) list of the input itemset.

These two mappings α and β are proved to induce a Galois connection between the
powerset of objects and the powerset of items. This Galois connection means that α
and β are dually adjoint, that is, $O \subseteq \beta(I) \Leftrightarrow I \subseteq \alpha(O)$ for a set of objects O and set
of items I. As a consequence of this property, these two mappings can be combined
in a pair of composite operators: $\alpha \cdot \beta$ which maps the powerset of items into itself,
and $\beta \cdot \alpha$ which maps the powerset of objects into itself. It follows immediately
by the theory that these two compositions are *closure operators* (i.e. they keep the
properties of monotonicity, extensivity and idempotency).

A *formal concept* in the context is a pair (O,I) of a set of objects $O \subseteq \mathcal{O}$ and a
set of items $I \subseteq \mathcal{I}$, where $\alpha(O) = I$ and $\beta(I) = O$; in this case O is called the extent
and I the intent of the concept. E.g., for data in Figure 2.1 one of the concepts is
$(\{1,2\},AC)$. For each concept (O,I), it follows that $\alpha \cdot \beta(I) = I$ and $\beta \cdot \alpha(O) = O$;
so, intents are *closed set of items* and extents *closed set of objects*. The concepts of a
context can be interpreted also from the geometrical point of view, they are maximal
rectangles of 1's in the binary table R representing the context. This idea is the basis
of other different works such as [62], where the maximal rectangles of 1's are called
tiles, and the goal is to extract all tiles with area over a given threshold.

From the set of formal concepts extracted from the context, we can draw a *lattice of concepts*: a Hasse diagram where each node is a concept, and there is an
edge between two nodes if and only if they are comparable and there is no other

Fig. 2.2 Example of a classical concept lattice

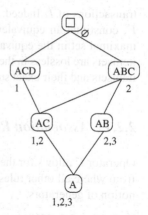

intermediate concept in the lattice. Indeed, this is simply a partial order organization of concepts where edges correspond to the standard inclusion of sets; then, the ascending paths in the lattice represent the subclass/superclass relation. The top of this lattice is represented by the unsatisfiable constant, thus meaning a set of items not included in any object (commonly also represented by the set of all attributes in the data).

A graphical representation of the lattice for data in Figure 2.1 can be followed from Figure 2.2: each node contains a closed itemset (intent) and it is labelled by the corresponding list of closed sets of objects (extent) linked by the Galois connection. So, each node corresponds to a formal concept. Observe also that each transaction of our input data is always a closed itemset, and the other closed itemsets in the lattice correspond to the intersection of subsets of input tuples. The theory ensures that the intersection of a number of extents (or intents) is always another extent (intent); thus, we have a closure system. This observation will be useful in next section, where we argue the connection of FCA with the classical propositional Horn theory.

2.2 The Closure System of Itemsets

In this section we want to focus on the closure operator working on the sets of items, that we name for short $\Gamma = \alpha \cdot \beta$. From the point of view of Γ, closed sets of items are those coinciding with their closure, that is, for $I \subseteq \mathscr{I}$, I is *closed* if $\Gamma(I) = I$; for example, we have that $\Gamma(AC) = AC$ for data in Figure 2.1.

From the practical point of view of data mining algorithms, such as in [18, 102, 103, 143], closed itemsets are maximal among those other itemsets occurring in the same transactions, i.e. $X \subseteq \mathscr{I}$ is *closed* when there is no other set $Y \subseteq \mathscr{I}$ such that $X \subset Y$ and $supp(X) = supp(Y)$, where $supp$ denotes the support of the itemset. Observe that the formalization of operator Γ captures exactly the algorithmic definition with support: $\Gamma(I)$ for an itemset $I \subseteq \mathscr{I}$, is the intersection of the set of objects where I is contained; this returns another set $I' \supseteq I$ occurring in the same transactions of I. Indeed, those itemsets having the same closure when operating Γ, constitute an equivalence class of sets with the same support. Then, only the maximal set in the equivalence class will be the closed one [19]. In theory, closed itemsets are lossless in the sense that they uniquely determine the set of all frequent itemsets and their exact support (c.f. [107] for more theoretical details).

2.2.1 Association Rules

Operator Γ allows for the construction of a base of nonredundant association rules from where all other rules can be derived. This construction is possible through the notion of generators.

Definition 2.1 ([3]). An association rule is a pair (G, Z), denoted $G \to Z$, where $G, Z \subseteq \mathscr{I}$ and $G \subseteq Z$.

When $\Gamma(G) = Z$ for an itemset $G \neq Z$ and G is minimal among all the candidates with closure equal to Z, we say that G is a *generator* of Z. For example, for the concept lattice in Figure 2.2 we have that B is a generator of AB because $\Gamma(B) = AB$ and B is minimal. Generators are presented under different names in the current literature: e.g. they receive the name of free sets in [25], or key patterns in [19], also, generators are extended to other condensed representations, like the disjunct-free itemsets [29], the non derivable itemsets [31], or the minimal k-free representations of frequent sets [32].

Under the FCA method, we are interested in implications of the form $G \to Z$, where G is a generator of Z. These turn out to be the particular case of association rules where no support condition is imposed but confidence is 1 (or 100%) [46, 70, 108, 102]. Such rules in this context are sometimes called *deterministic association rules*. For example, from the concept lattice in Figure 2.2 we could generate a deterministic association rule $B \to AB$, since B is a generator of the closed set AB as mentioned above.

As detailed in the introduction, the foundations of deterministic association rules in FCA were initially studied by Duquenne and Guigues in [46]. Later, the work by Luxenburger in [93] proposed a way to derive association rules whose confidence was less than 1; these would correspond to the *nondeterministic association rules*. Different algorithmic contributions to nondeterministic rules, where the notion of support was incorporated, are [18, 102, 120, 144, 137, 141]. Other works such as [25] derived the idea of δ-strong rules (the rule is violated in no more than δ rows, and δ is supposed to have a small value, so that the rule has very few exceptions).

The interest of the deterministic point of view given by [46] is that the association rules can be studied under the framework of Horn logic. The binary relation of the formal context can be naturally viewed also as sets of models (0/1 assignments to n propositional variables). Then, deterministic association rules can be seen as propositional logic formulas capturing information contained in a set of models. Practically effective approaches to find such logical formulas have been proposed in the field of Knowledge Compilation ([30, 113]): among them, a prominent basic process is to "compile" the list of satisfying models into a tractable set of Horn clauses ([80, 113]). The next section gives an overview of these results.

2.3 Classical Propositional Horn Logic

Assume a standard propositional logic language with propositional variables. The number of variables is finite and we denote by $\mathcal{V}$ the set of all variables; we could alternatively use an infinite set of variables provided that the propositional issues corresponding to a fixed dataset only involve finitely many of them. A literal is either a propositional variable, called a positive literal, or its negation, called a negative literal. A clause is a disjunction of literals and it can be seen simply as the set of the literals it contains. A clause is *Horn* if and only if it contains at most one positive literal. Horn clauses with a positive literal are called *definite*, and can be written as $H \to v$ where H is a conjunction of positive literals that were negative in the clause, whereas v is the single positive literal in the clause. Horn clauses without positive literals are called *nondefinite*, and can be written similarly as $H \to \square$, where $\square$ expresses unsatisfiability. A Horn formula is a conjunction of Horn clauses.

A *model* is a complete truth assignment, i.e. a mapping from the variables to $\{0, 1\}$. We denote by $m(v)$ the value that the model m assigns to the variable v. The intersection of two models is the bitwise conjunction returning another model. A model satisfies a formula if the formula evaluates to true in the model. The universe of all models is denoted by $\mathcal{M}$.

A theory is a set of models. A theory is Horn if there is a Horn formula which axiomatizes it, in the sense that it is satisfied exactly by the models in the theory. When a theory contains another we say that the first is an upper bound for the second; for instance, by removing clauses from a Horn formula we get a larger or equal Horn theory. The following is known (c.f. [37], or works such as e.g. [80]):

Theorem 2.1. *Given a propositional theory of models M, there is exactly one minimal Horn theory containing it. Semantically, it contains all the models that are intersections of models of M. Syntactically, it can be described by the conjunction of all Horn clauses satisfied by all models from the theory.*

The theory obtained in this way is called sometimes the *empirical Horn approximation* of the original theory. Clearly, then, a theory is Horn if and only if it is actually *closed under intersection*, so that it coincides with its empirical Horn approximation.

2.3.1 Closures and Horn Clauses

The propositional Horn logic framework allows us to cast our reasoning in terms of closure operators. Naturally, the set of objects $\mathcal{O}$ of the formal context can be seen as a set of models. It turns out that it is possible to exactly characterize the set of deterministic association rules in terms of propositional logic: we can associate a propositional variable to each item, and each association rule becomes a conjunction of Horn clauses. Then:

Theorem 2.2. *[14] Given a set of transactions, the conjunction of all the deterministic association rules defines exactly the empirical Horn approximation of the theory formed by the given tuples.*

So, the theorem determines that the empirical Horn approximation of a set of models can be computed with the FCA method of constructing deterministic association rules, that is, constructing the closed sets of attributes and identifying minimal generators for each closed set. A similar lattice theoretic approach was already used in works such as [43, 44] to analyze functional dependencies in the data.

2.4 Learning with Lattices

The resulting graph-based system produced by FCA provides interesting insights into different learning tasks. One application is to understand the Galois lattice as a basic classification of input objects: the closed itemsets are the discriminating features, and the closed set of objects represent the different clusters. In other words, each formal concept is indeed a group of input objects classified in the same class. In this line of research, the work of [61] extends the notion of closed sets with labels, to be used in the construction of classifiers based on association rules. Previous to that, the work in [33] presented an approach to conceptual clustering, named GALOIS. GALOIS can be used for class discovery and class prediction, and it represents and updates all possible classes in a restricted concept space. Another work is [111], presenting an algorithm named Rulelearner where the Galois lattice is used as an explicit guide through the rule space so as to induce classification rules; the author shows that this algorithm is also capable of learning decision lists. On the other hand, in [99] the authors describe an instance-based learning system over lattice theory called IGLUE, that improves both the complexity and accuracy of other lattice-based learning systems.

In other different applications the lattice has proved to be also very effective: for example as a methodology for a graph-based learning system in [65], or in the information retrieval of documents [34], or as a formalization of top-down induction systems in [50], among many others. Other different applications in this binary context are documented in [35].

One of the advantages of having a properly formalized closure system is that the search space to identify a valid hypothesis is smaller, thus it can be used to potentially improve the efficiency of learning algorithms. As pointed out in the introduction, many algorithms have been proposed that search for closed itemsets in binary data. Those closed itemsets can be organized in a Hasse diagram, Galois lattice, explained in this chapter.

Although the Galois lattice has proved to be a powerful tool to formalize a reduced hypothesis space to be used in classification tasks, an important limitation is the classical propositional description of the examples. This is an important drawback, specially in the field of knowledge discovery where objects to be mined usually exhibit a complex structure not always represented as binary contexts. From this point of view, there is already some research focusing on the extension of the description language of closed patterns into sequences [135, 125, 131, 132, 38, 109], episodes [146, 122, 106, 105], trees [6, 84, 13], graphs [92, 136, 23, 24], or other relevant works extending the representation to multi-relational queries, such as e.g. [110, 118, 60]. Most of these approaches are mainly algorithmic, focusing on finding a frequency-based closure in a efficient way. This manuscript will focus on the formal characterization of closed patterns. In particular we will focus on the closure of sequential patterns and their organization in a lattice of closed concepts. From this combinatorial object, we will introduce the notion of closed partial orders and the closure of structured data without cycles.

Chapter 3
Lattice Theory for Sequences

The goal of this chapter is to use FCA theory to formalize a new closure system that characterizes sequential data. Since we are not dealing with the classical unordered context of the preliminaries, setting all the conditions for the new Galois connection is not a trivial task. To start with, it departs from the unordered case in the very definition of intersection; whereas we saw in the last chapter that the intersection of two itemsets is another itemset, the intersection of two or more sequences is not necessarily a single sequence. Let us consider the following definition.

Definition 3.1. The **intersection** of a collection of sequences $s_1, \ldots, s_n \in \mathscr{S}$, denoted $s_1 \cap \ldots \cap s_n$, is the set of maximal subsequences contained in all the s_i.

Due to the maximality condition (defined after Definition 1.1), the following property is clear.

Proposition 3.1. *If $s \subseteq s_1$, $s \subseteq s_2, \ldots, s \subseteq s_n$ then there is some s' in $s_1 \cap \ldots \cap s_n$ such that $s \subseteq s'$.*

For example, the intersection of $s = \langle (AD)(C)(B) \rangle$ and $s' = \langle (A)(B)(D)(C) \rangle$ is the set of sequences $\{\langle (A)(C) \rangle, \langle (A)(B) \rangle, \langle (D)(C) \rangle\}$: all of them are contained in s and s' and among those having this property they are maximal; all other common subsequences are not maximal since they can be extended to one of these. The maximality condition of the intersection discards redundant information since the presence of, e.g. $\langle (A)(B) \rangle$, already informs of the presence of the itemsets (A) and (B) individually. Indeed, this notion of intersection formalizes one of the intuitions mentioned in the first chapter: we noticed that different closed sequential patterns, as defined by CloSpan, may coexist together in exactly the same input transactions. Here, the definition of the intersection operation naturally models the occurrence of maximal sequential patterns in the same set of input ordered data.

G.C. Garriga: *Formal Methods for Mining Structured Objects*, SCI 475, pp. 21–37.
DOI: 10.1007/978-3-642-36681-9_3   © Springer-Verlag Berlin Heidelberg 2013

3.1 The Ordered Context

The notion of ordered context is meant to provide an alternative view of our set of sequences as the objects needed by the FCA theory. As it was introduced in the last chapter, the classical formal context corresponds to a binary relationship between the set of objects and the items they contain. Yet, when dealing with ordered data each object represents a sequence in $\mathscr{D}$, and thus, items in each object are partially ordered through time. Formally:

Definition 3.2. We define our **ordered context** as a tuple $(R, \mathscr{O}, \mathscr{I})$, where $R \subseteq \mathscr{O} \times \mathscr{I} \times \mathbb{N}$. The elements of $\mathscr{O}$ correspond to the set of objects in the new context and those of $\mathscr{I}$ the items. For an entry $(o, i, k) \in R$ we read "the item i occurs at the position k in the object o"; so, k represents the *order* of occurrence of i with respect to the other attributes in the same object.

Like the one-valued contexts treated in the preliminaries, ordered contexts can be also represented by a cross table, the rows of which are labelled by the objects and the columns labelled by the items. A simple example of a set of input transactions and its associated context is found in Figure 3.1: each input sequence in $\mathscr{D}$ can be represented as an object, so that the entry in row o_j and column i corresponds to the order/s of the item i in the equivalent input sequence d_j.

Seq id	Input sequences
d_1	$\langle (AE)(C)(D)(A) \rangle$
d_2	$\langle (D)(ABE)(F)(BCD) \rangle$
d_3	$\langle (D)(A)(B)(F) \rangle$

(a) Collection of data $\mathscr{D}$

Objects	A	B	C	D	E	F
o_1	1,4		2	3	1	
o_2	2	2,4	4	1,4	2	3
o_3	2	3		1		4

(b) Ordered context for $\mathscr{D}$

Fig. 3.1 Example of ordered data $\mathscr{D}$ and its ordered context

Naturally, the ordered context for a set of sequences is relevant to this work to see objects $\mathscr{O}$ and input sequences $\mathscr{D}$ as equivalent sets. Notation provided by the objects will be used under the FCA perspective, whereas the equivalent set of input sequences belongs to the database terminology.

3.2 A New Closure Operator

To motivate the need to characterize a specific closure operator for the ordered context, we first analyze other possibilities suggested by the classical theory. A first idea is to transform our set of sequences into a single normalized table, as [51] proposed to do for the multi-valued contexts. This transformation process applies a "conceptual scaling" that consists on expanding each original attribute with the labels of any of its possible values; then, with the new set of attributes, the context is translated into a binary one. It is easy to see that for the ordered context this scaling would not work: the new set of attributes would correspond to extending the items with their different positions in the objects; yet, items do not always occur in the same position in all the objects, which can make the intersection fail. For instance, a subsequence $\langle(A)(B)\rangle$ can occur in positions $\langle(A,1),(B,2)\ldots\rangle$ in one input sequence, but in positions $\langle\ldots(A,3),(B,5)\ldots\rangle$ in another input sequence. The intersection of these two sequences should clearly return at least the sequence $\langle(A)(B)\rangle$; however, when extending the items into the labels $A3$, $A1$ and $B5$, $B2$, we would have that they correspond to different attributes, so that they never intersect.

Another option to get a closure operator on the ordered context would be to treat the data as an ordinal context, presented also in [51]. But again, ordinal contexts do not serve to express the order we want on the items of an object. In fact, an ordinal context is just a specific case of multivalued contexts describing an intra-attribute order, yet, we would better need an inter-attribute order for our ordered context. Therefore, we decide to propose our own characterization of the properly defined closure operator.

3.2.1 Defining the Galois Connection

We define the following two *derivation operators*: ϕ and ψ. To follow (approximately) the standard terminology of FCA we denote objects with $\mathcal{O} = \{o_1,\ldots,o_n\}$ (or, to simplify, as the natural numbers from 1 to n, where n is the size of $\mathcal{D}$) and use capital letters O, O' for subsets of $\mathcal{O}$. Note that each object o identifies exactly one transaction d_o of $\mathcal{D}$; however, it could be the case that the input sequence d_o itself does not identify o since there could be repeated sequences in the input database.

Given that in the ordered context the intersection of a set of input sequences may result in more than one sequence, we propose a first operator mapping a set of objects into a set of sequences, i.e. $\phi : 2^{\mathcal{O}} \longrightarrow 2^{\mathcal{S}}$. For consistency with this operator, the dual mapping is defined as the inverse $\psi : 2^{\mathcal{S}} \longrightarrow 2^{\mathcal{O}}$.

- For a set $O \subseteq \mathcal{O}$ of objects we define,

$$\phi(O) = \{s \in \mathcal{S} \,|\, s \text{ is maximally contained in } d_o, \text{ for all } o \in O\}.$$

The mapping $\phi(O)$ returns the set of sequences common to *all* the objects in O. In fact, $\phi(O)$ is nothing but the intersection of the input sequences in $\mathcal{D}$ identified by O.

- Correspondingly, for a set $S \subseteq \mathcal{S}$ of sequences we define,

$$\psi(S) = \{o \in \mathcal{O} \,|\, s \subseteq d_o, \text{for all } s \in S\}.$$

Function $\psi(S)$ returns the set of (indices of) input sequences that include *all* the sequences in S, that is, for the case of single sequences, $\psi(\{s\}) = tid(s)$, and, in general, $\psi(S)$ returns the transaction identifier list for the set of sequences S in $\mathcal{D}$. Of course $\psi(S)$ could be the empty set.

For example, for the context in Figure 3.1 we have that $\phi(\{1,3\}) = \{\langle (D)(A) \rangle\}$, and $\psi(\{\langle (AE)(D) \rangle, \langle (AE)(C) \rangle\}) = \{1,2\}$. Sets of objects can be ordered by the standard inclusion $\subseteq$ (indeed, they simply correspond to indexes), as in the binary context, while sets of sequences will be ordered by the properly defined relation $\preceq$.

Definition 3.3. We say that a set of sequences S is **more general** than another set of sequences S', denoted by $S \preceq S'$, if and only if for all $s \in S$, there exists $s' \in S'$ such that $s \subseteq s'$. Then S' is also said to be **more specific** than S.

According to the definition, the set of sequences $\{\langle (A) \rangle, \langle (B) \rangle, \langle (C) \rangle\}$ is more general than the set $\{\langle (B)(A) \rangle, \langle (C)(A) \rangle\}$. In fact this relation is not an ordering, but only a preorder; however, it will work as an order on those cases where no sequence in a set is a subsequence of another sequence of the same set, and this will be the case of interest (and the reason why the notion of intersection is restricted to maximal subsequences). Finally, note that $s \subseteq s'$ if and only if $\{s\} \preceq \{s'\}$, and that $S \preceq \{s'\}$ means that all the sequences in S are subsequences of s'.

Now we have the following property, corresponding to the properties of Galois connections (although formally they must be used on orderings and we are only in the presence of a preorder, so that in its fully formal definition this pair is not completely a Galois connection).

Proposition 3.2. *For sets of objects $O, O' \subseteq \mathscr{O}$, and sets of sequences $S, S' \subseteq \mathscr{S}$, the following properties hold:*

1)	$O \subseteq O' \Rightarrow \phi(O') \preceq \phi(O)$	1')	$S \preceq S' \Rightarrow \psi(S') \subseteq \psi(S)$
2)	$O \subseteq \psi(\phi(O))$	2')	$S \preceq \phi(\psi(S))$

Proof. Each one of these properties can be proved as follows:

1) For all $s' \in \phi(O')$ we have that, for all $o' \in O'$, s' is contained in o', that is, in particular s' is contained in o for all $o \in O$, if $O \subseteq O'$, and therefore (Proposition 3.1) there exists $s \in \phi(O)$ s.t. $s' \subseteq s$, which means $\phi(O') \preceq \phi(O)$.

2) For all $o \in O$ we have that, s is contained in o for all $s \in \phi(O)$, and thus $o \in \psi(\phi(O))$.

1') For all $o' \in \psi(S')$ we have that, for all $s' \in S'$, s' is contained in o', that is, in particular s is contained in o' for all $s \in S$, if $S \preceq S'$, and thus $o' \in \psi(S)$, which means $\psi(S') \subseteq \psi(S)$.

2') For all $s \in S$ we have that, s is contained in o for all $o \in \psi(S)$, and thus there exists $s' \in \phi(\psi(S))$ s.t. $s \subseteq s'$, which implies $S \preceq \phi(\psi(S))$. $\square$

The following property readily follows from Proposition 3.2: the two derivation operators ϕ and ψ are dually adjoint.

Proposition 3.3. $O \subseteq \psi(S) \Leftrightarrow S \preceq \phi(O)$.

From these properties, we can obtain two closure systems that are dually isomorphic to each other: one on sets of objects, obtained from the composition $\widehat{\Delta} = \psi \cdot \phi$, and another on sets of sequences, from $\Delta = \phi \cdot \psi$. In fact, Δ is our operator of interest, and works as follows: the closure $\Delta(S)$ of a set of sequences $S \in \mathscr{S}$, includes all the maximal sequences that are present in all objects having all sequences in S; that is, the intersection of all those input sequences $d \in \mathscr{D}$ such that $S \preceq \{d\}$. Taking the example from Figure 1.1, we have that $\Delta(\{\langle (D) \rangle\})$ corresponds to the intersection of input sequences d_1, d_2 and d_3, i.e. the equivalent input sequences to objects o_1, o_2 and o_3 of $\phi(\{\langle (D) \rangle\}) = \{1,2,3\}$. Then, $\Delta(\{\langle (D) \rangle\}) = \{\langle (D)(A) \rangle\}$.

The following proposition follows directly from this formalization.

Proposition 3.4. *Compositions* $\widehat{\Delta} = \psi \cdot \phi$ *and* $\Delta = \phi \cdot \psi$ *are closure operators.*

Proof. According to [42] or [51], to show that Δ is a closure operator we need to prove: monotonicity: $S \preceq S'$ implies $\Delta(S) \preceq \Delta(S')$; extensivity: $S \preceq \Delta(S)$; and idempotency: $\Delta(\Delta(S)) = \Delta(S)$. These properties follow immediately from the facts in Proposition 3.2.

- Monotonicity: $S \preceq S'$ by 1') yields to $\psi(S') \subseteq \psi(S)$, and by 1) we get $\phi(\psi(S)) \subseteq \phi(\psi(S'))$.
- Extensivity: follows directly from 2').
- Idempotency: by property 2') we have that $\phi(\psi(S)) \preceq \phi(\psi(\phi(\psi(S))))$, thus $\Delta(S) \preceq \Delta(\Delta(S))$; on the other hand, with $O := \psi(S)$ we obtain by property 2) that $\psi(S) \subseteq \psi(\phi(\psi(S)))$, and then property 1) yields to $\phi(\psi(\phi(\psi(S)))) \preceq \phi(\psi(S))$, thus $\Delta(\Delta(S)) \preceq \Delta(S)$. Equality follows from the fact that the closure operator, by definition of ϕ, always gives a set where no sequence is a proper subsequence of another.

Symmetrically, the dual operator $\widehat{\Delta}$ can be proved to be a closure operator on the universe of objects. □

As customary, *closed sets of sequences* are those coinciding with their closure, that is, $\Delta(S) = S$. A simple, but necessary observation is the following proposition.

Proposition 3.5. *All sequences in a closed set are maximal in it w.r.t.* $\subseteq$.

Proof. Since S is a closed set of sequences we have that $S = \Delta(S) = \phi(\psi(S))$. The definition of ϕ ensures that each $s \in S$ is a maximal sequence w.r.t. $\subseteq$, common to all of $\psi(S)$. □

Note that $\{d\}$ is closed for each individual d of our database $\mathscr{D}$. In fact, $\phi(\{d\})$ may return more than one object (if d is a subsequence of other input sequences), but because at least one of those objects corresponds exactly to d, we have that their intersection returns always d again; thus, $\psi(\phi(\{d\}))$ is d.

3.3 Formal Concepts and the Concept Lattice

Definition 3.4. A **formal concept** of the ordered context is a pair (O, S) where $O \subseteq \mathcal{O}$, $S \subseteq \mathcal{S}$, and $\phi(O) = S$ and $\psi(S) = O$. Then O is the extent and S is the intent of the concept, and both sets are said to be linked by the Galois connection. These concepts are also called closed since we have that $\Delta(S) = S$ and $\widehat{\Delta}(O) = O$.

For every set $O \subseteq \mathcal{O}$, $\phi(O)$ is the intent of some concept, since $(\psi(\phi(O)), \phi(O))$ will always be a concept. Consequently, a set $O \subseteq \mathcal{O}$ is an extent if and only if it is closed, $O = \psi(\phi(O)) = \widehat{\Delta}(O)$. The same applies to intents, corresponding to closed set of sequences, that is $S = \phi(\psi(S)) = \Delta(S)$. Notice that the difference of this formal concept from the classical one-valued concept relies on the intent. Here, it consists on a *set* of *maximal* sequences common to the objects of the extent.

Given the context of Figure 3.1, the set of all closed concepts are shown in Figure 3.2. Each concept corresponds to a closed set of sequences, and the associated extent is a list of those objects linked by the Galois connection. These sets form, at the same time, the dual system of closed sets of objects.

Fig. 3.2 Closed concepts from data in Figure 3.1

Extent	Intent
$\{1,2,3\}$	$\{\langle (D)(A) \rangle\}$
$\{2,3\}$	$\{\langle (D)(A)(B) \rangle, \langle (D)(A)(F) \rangle, \langle (D)(B)(F) \rangle\}$
$\{1,2\}$	$\{\langle (AE)(C) \rangle, \langle (AE)(D) \rangle, \langle (D)(A) \rangle\}$
$\{1\}$	$\{\langle (AE)(C)(D)(A) \rangle\}$
$\{2\}$	$\{\langle (D)(ABE)(F)(BCD) \rangle\}$
$\{3\}$	$\{\langle (D)(A)(B)(F) \rangle\}$

Interestingly enough, the basic properties of a closure system still hold on our formal concepts: the union of extents does not result necessarily in another extent, but the intersection of any number of extents/intents is always another extent/intent.

Proposition 3.6. *The intersection of intents/extents is another intent/extent.*

Here, by *intersecting two set of sequences*, $S_1 \cap S_2$, we understand the set of maximal sequences resulting from the cross intersection of $s_1 \cap s_2$, for all $s_1 \in S_1$ and all $s_2 \in S_2$.

Proof. Let S_1 and S_2 be two closed sets of sequences, that is we have that $\Delta(S_1) = S_1$ and $\Delta(S_2) = S_2$. By monotonicity of the closure operator Δ we get: first, $S_1 \cap S_2 \preceq S_1$ implies $\Delta(S_1 \cap S_2) \preceq \Delta(S_1)$, that is, $\Delta(S_1 \cap S_2) \preceq S_1$; and second, $S_1 \cap S_2 \preceq S_2$ implies $\Delta(S_1 \cap S_2) \preceq \Delta(S_2)$, that is, $\Delta(S_1 \cap S_2) \preceq S_2$. Therefore, we have that $\Delta(S_1 \cap S_2) \preceq S_1 \cap S_2$. On the other hand, the extensivity of operator Δ ensures that $S_1 \cap S_2 \preceq \Delta(S_1 \cap S_2)$, so that (resorting again to the maximality condition) we necessarily have that $\Delta(S_1 \cap S_2) = S_1 \cap S_2$.   □

From the set of all our formal concepts we can construct the concept lattice of the ordered context. The concept lattice is a Hasse diagram organizing the concepts by the following relation.

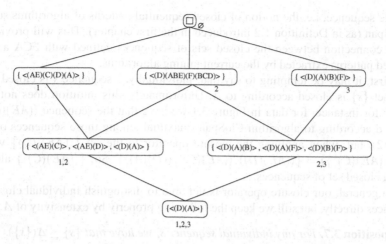

Fig. 3.3 Example of a concept lattice for the database of Figure 3.1

Definition 3.5. If (O, S) and (O', S') are concepts of a context, we say that (O, S) is a subconcept of (O', S') if $O' \subseteq O$ (equivalent to $S \preceq S'$).

In Figure 3.3 we show the representation of the concepts of Figure 3.2: each node corresponds to an intent of a concept, which is labelled by its extent; edges in the lattice correspond to the order between concepts (set-theoretic inclusion downwards by the extents, and $\preceq$-inclusion upwards by the intents).

The set of sequences contained in all the input sequences will be called the *bottom* of the lattice; in most cases it will happen to be a trivial, somewhat artificial, element containing only the empty sequence. Similarly, we can also add an artificial set of sequences not contained in any object; this will represent the *top* of the concept lattice. In the example from Figure 3.3 an artificial top not belonging to any object is denoted by the unsatisfiable boolean constant $\square$. This artificial element is added to the lattice just to the effect of our later arguments.

Finally, we say that a closed set of sequences S' is an *immediate predecessor* (or also proper predecessor) of another closed set of sequences S s.t. $S \neq S'$ if $S' \preceq S$ and no other closed set S'' exists in the lattice with $S' \preceq S'' \preceq S$. For example in Figure 3.3: $\{\langle(AE)(C)\rangle, \langle(AE)(D)\rangle, \langle(D)(A)\rangle\}$ and $\{\langle(D)(A)(B)\rangle, \langle(D)(A)(F)\rangle, \langle(D)(B)(F)\rangle\}$ are immediate predecessors of the node $\{\langle(D)(ABE)(F)(BCD)\rangle\}$. Similarly, successors are found via the ascending paths of each node, and immediate successors are located just one level upwards.

3.4 Closed Sequential Patterns in the Lattice

Closure operator Δ can be used only on a set of sequences, not single individual sequences. In this section we want to derive from our model the closure of a single sequence, i.e. the notion of closed sequential patterns of algorithms such as CloSpan (as in Definition 1.2 introduced in the first chapter). This will provide the final connection between the closed sets of sequences defined with FCA and the closed patterns extracted by the current mining algorithms.

First, it may be tempting to say that, individually, a sequence s is closed when the set $\{s\}$ is closed according to Δ. Unfortunately, this intuition does not work here; for instance, for data in Figure 3.1 we have that the sequence $\langle (AE)(C) \rangle$ is closed according to algorithm CloSpan (maximal among those sequences of support 2), but, when applying the closure operator to the set $\{\langle (AE)(C) \rangle\}$ we get $\Delta(\{\langle (AE)(C) \rangle\}) = \{\langle (AE)(D) \rangle, \langle (AE)(C) \rangle, \langle (D)(A) \rangle\}$. So, $\{\langle (AE)(C) \rangle\}$ alone is not a closed set of sequences.

In general, our closure operator is not able to distinguish individual closed sequences directly, but still we keep the following property by extensivity of Δ.

Proposition 3.7. *For any individual sequence s, we have that $\{s\} \preceq \Delta(\{s\})$.*

So, if we try to close a single sequence s with $\Delta(\{s\})$, we get a closed set of sequences where at least one of them is a supersequence of s. This property naturally leads to the following definition.

Definition 3.6. A single sequence s is **stable** under Δ if $s \in \Delta(\{s\})$.

Usually Δ is clear from the context and thus, it is omitted from the definition, so that we simply speak of stable sequences. This notion of stability gathers all those *maximal* sequences occurring in a *maximal* set of objects of the context; that is, individual sequences belonging to closed sets of sequences. In other words, a sequence s is stable for a set of objects O when we cannot add any other object to the set O without losing this s in the set of sequences obtained with $\phi(O)$.

Notice that not all the sequences are stable, for example, sequence $\langle (A)(C) \rangle$ from the data of Figure 3.1 is not stable because we have that: $\Delta(\{\langle (A)(C) \rangle\}) = \{\langle (AE)(C) \rangle, \langle (AE)(D) \rangle, \langle (D)(A) \rangle\}$. Indeed, just the closed sequential patterns can be characterized as stable sequences, i.e. we can prove that both closed sequential patterns, extracted from data $\mathscr{D}$ by CloSpan, and stable sequences, defined via the closure operator Δ, are equivalent.

Lemma 3.1. *The set of stable sequences coincides with the set of closed sequential patterns.*

Proof. We will prove both directions of the equality: that any stable sequence is also a closed sequence, and that any closed sequence is in fact a stable sequence.

⇒/ If a sequence s is stable then $s \in \phi(\psi(\{s\}))$, which leads to: first, $|O| = supp(s)$ where $O = \psi(\{s\})$ and $|O|$ defines the cardinality of set O; and, second, s is *maximal* by Proposition 3.5. So, we can rewrite that there is no other sequence s' s.t. $s \subset s'$ and s' is also contained in objects $o \in O$, that is, there is no supersequence of s with the same support. Thus, s is a closed sequence according to Definition 1.2.

⇐/ Let s be a closed sequence, and let $D \in \mathscr{D}$ be the set of all input sequences where this s is included, that is $|D| = supp(s)$. Because s is a closed sequential pattern, there is no s' s.t. $s \subset s'$ and $supp(s') = supp(s)$, or in other words, s is max-imal contained in the set of transactions D. Thus, $s \in \phi(O)$, where O is formed by the tids of D in $\mathscr{D}$, and we necessarily have that $O = \psi(\{s\})$. Then, we get that $s \in \phi(\psi(\{s\}))$; so, s is stable. $\qquad\qquad\qquad\qquad\qquad\qquad\qquad\qquad\qquad\qquad\qquad\qquad\qquad\qquad\qquad$ $\square$

After proving that both sets of patterns are the same, we will rather name these specific sequences as stable sequences instead of closed sequential patterns. In this way, we will avoid the confusion of the term "closed", which we want to use to refer to the closure operator Δ that only applies to sets of sequences.

Finally, to complete the characterization, we are interested in proving that the notion of stability, thus the notion of closed sequential patterns, is included in the concepts of our lattice. In other words, we are looking for a way to construct our clo-sure system from the stable sequences, which we have just shown can be mined by proper algorithms. The next result proves that the closed sets of sequences, thus the intents of our lattice, correspond to a set of raw stable sequences grouped together. This is the first step towards the lattice construction.

Theorem 3.1. *Let S be a closed set of sequences, then for all $s \in S$, s is stable.*

Proof. Let $\psi(S) = O$ and $\phi(O) = S$. For each single sequence $s \in S$ we examine the result of $\Delta(\{s\}) = \phi(\psi(\{s\}))$ to prove stability of s. Let $\psi(\{s\}) = O'$, so that:

- If $O' = O$, then $\phi(\psi(\{s\})) = S$, and since $s \in S$ we have that s is stable.
- If $O \subset O'$ (since it may well be that $\{s\}$ alone was contained in more objects than S, because $\{s\} \preceq S$), then by definition of Galois connection we have that $O \subseteq O'$ implies $\phi(O') \preceq \phi(O)$. Thus, $\phi(O') \preceq S$, which can be rewritten as $\phi(\psi\{s\}) \preceq S$. Since $s \in S$, then we get along with Proposition 3.7 the following relation on the sets $\{s\} \preceq \phi(\psi(\{s\})) \preceq S$. However, we know that s is *maximal* in S by Proposition 3.5, so that there is no $s' \in S$ s.t. $s \subset s'$; therefore, $s \in \phi(\psi(\{s\})) = \Delta(\{s\})$ and we conclude that s is stable.
- Finally, it cannot occur that $O' \subset O$, since $\{s\} \preceq S$ and so $\{s\}$ has to be contained in at least the same objects where S is contained. $\qquad\qquad\qquad\qquad$ $\square$

The table of Figure 3.4 shows the list of stable sequences that algorithms such as CloSpan would identify from the database presented in Figure 3.1 (without a mini-mum support condition). As stated by the theorem, each one of the stable sequences corresponds to an individual sequence in the intents, and vice versa. We also observe from the figure that some stable sequences are contained in more than one node of the lattice. The reason is that closed sets of sequences in the intents of each concept must keep the order given by $\preceq$, and so, the correspondence between closed sets and stable sequences may not be one to one.

As a consequence of characterizing our lattice in terms of stable sequences, we have that the construction of the lattice can be done by conveniently organizing the raw stable sequences into closed sets of sequences. From the more practical point of view, such a construction will avoid the intersection of input sequences implicit in the calculation of Δ. The next section gives the details of such organization.

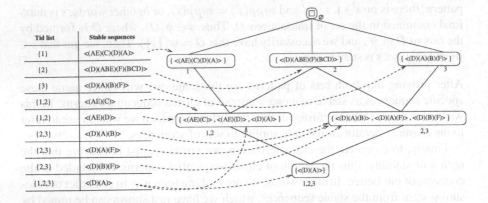

Tid list	Stable sequences
{1}	<(AE)(C)(D)(A)>
{2}	<(D)(ABE)(F)(BCD)>
{3}	<(D)(A)(B)(F)>
{1,2}	<(AE)(C)>
{1,2}	<(AE)(D)>
{2,3}	<(D)(A)(B)>
{2,3}	<(D)(A)(F)>
{2,3}	<(D)(B)(F)>
{1,2,3}	<(D)(A)>

Fig. 3.4 All the stable sequences derived from data in Figure 3.1 and their correspondence with the lattice

3.5 Lattice Construction

Organizing raw stable sequences into closed sets of sequences is not a direct task. First, some of these stable sequences may belong to several intents (e.g. the sequence $\langle (D)(A) \rangle$ of Figure 3.4), and second, when constructing the lattice we still need to ensure that the final system is closed. To characterize such construction it is necessary to take into account two facts: first, for a formal concept (O, S), the set S contains a set of maximal stable sequences whose tid list is at least O, i.e. $O \subseteq tid(s)$ for all $s \in S$; and second, the final intents of the lattice must be organized by means of the order $\preceq$.

We start by setting some basic operations to ease the formalization of the final result: let S and S' be two sets of sequences, then we need to define the following two operations,

- $S \ominus S' = \{s | s \in S \text{ but } \{s\} \not\preceq S'\}$, that returns all the sequences in S not included in any one of the sequences of S'.
- $\max\{S\} = \{s | s \in S, \text{ and there does not exist } s' \in S \text{ s.t. } s \subset s'\}$, that keeps the maximal sequences w.r.t. $\subset$ in a set.

The main property formalizing the construction of each node of the lattice by means of its proper predecessors is expressed in Theorem 3.2. This result will lead to the bottom-up construction of the final closure system once we have all the stable sequences of CloSpan. The specific algorithmic details following from this result will be provided in the next subsection.

Theorem 3.2. *Let (O, S) be a formal concept, and let $\{(O_i, S_i)\}$ indexed by i, be its family of immediate predecessors in the lattice. Then, 1/ $O = \bigcap O_i$ if $i > 1$, and 2/ let $\tilde{S} = \{s | s \text{ is stable with } tid(s) = O\}$, then $S = \tilde{S} \cup \max\{\bigcup(S_i \ominus \tilde{S})\}$.*

Note that usually the index i of intersections and unions is clear from the context and we remove it for simplicity. So, when writing $O = \bigcap O_i$, we actually mean $O = \bigcap_{i=1}^{n} O_i$.

Proof. To prove the first part of the theorem, we must consider that the dual set of objects is also a closure system ordered by set-theoretic inclusion downwards. This means that $O \subseteq O_i$ for each closed set of objects of its immediate predecessors indexed by i. Indeed, in set theory this implies that $O \subseteq \bigcap O_i$. On the other hand, we have that each O_i is defined as an immediate predecessor of O, so that we necessarily have $O = \bigcap O_i$ (otherwise there would exist a different formal concept between these two and O_i would not be an immediate predecessor). Notice that when we have only one immediate predecessor (i.e. index $i = 1$), then $O \subseteq O_1$.

To prove the second part we do as follows. We want S to be a closed set of sequences, so that $S = \phi(O)$ with $O = \psi(S)$. Our first observation is that every stable sequence s, maximal by construction, with $tid(s) = O$, will belong to $\phi(O)$ by definition, so that s must belong to S. This justifies that S is at least initiated with $\tilde{S}$ in the theorem. At the same time, it follows from the FCA theory that $O \subset O_i$ and $S_i \preceq S$ for each immediate predecessor (O_i, S_i). That is, for all sequences $s' \in S_i$, we have that s' is contained as a maximal sequence in all objects O_i, and therefore, s' is contained also in all objects O, although it might not be maximal there. Then, to keep the final proper order $S_i \preceq S$, all the sequences $s' \in S_i$ such that $\{s'\} \not\preceq \tilde{S}$, should be contained also in S. However, by definition, only the maximal of those sequences in $s' \in S_i$ will belong to $\phi(O)$. Hence, from all the differences $\bigcup S_i \ominus S$, we keep only the maximal ones. This justifies that $\tilde{S} \cup \max\{\bigcup(S_i \ominus \tilde{S})\} \subseteq S$.

The reverse inclusion of the second part of the theorem can be argued as follows. By construction, each $s \in S$ satisfies $\psi(\{s\}) \supseteq O$; then, either $tid(s) = O$ so that s belongs to $\tilde{S}$, or $O \subset tid(s)$. In this latter case, because s is a stable sequence (Theorem 3.1), we have that $s \in S'$ where $S' = \phi(\psi(\{s\}))$. It follows from the FCA theory that S' is a closed set of sequences in our lattice s.t. $S' \preceq S$; then, s belongs to a predecessor of S, named S'. Moreover, because the lattice is ordered by a subconcept relationship (i.e. $\preceq$-inclusion upwards), s will be included in some immediate predecessor of S, and moreover, it will be a maximal sequence from the set of all sequences in the predecessors. Therefore, we have that $S = \tilde{S} \cup \max\{\bigcup(S_i \ominus \tilde{S})\}$, as stated by the theorem.   □

Notice that the last result includes also those cases where $\tilde{S} = \emptyset$, so that the closed set of sequences is simply defined as $S = \max\{\bigcup(S_i)\}$. These nodes are necessary to ensure that the final system is closed.

After Theorem 3.2 we can naturally proceed to the bottom-up construction of the lattice out of the output of CloSpan, by conveniently grouping the mined stable sequences: for a closed set of objects O, the set of stable sequences in $\tilde{S}$ serves as an initialization to construct the intent S linked to O through the Galois connection;

Seq id	Input sequences
d_1	$\langle (C)(B)(A) \rangle$
d_2	$\langle (B)(C)(A) \rangle$
d_3	$\langle (C)(A) \rangle$
d_4	$\langle (B)(A) \rangle$

(a) Sequential data

	A	B	C
o_1	4	3	2
o_2	3	1	2
o_3	2		1
o_4	2	1	

(b) Ordered context

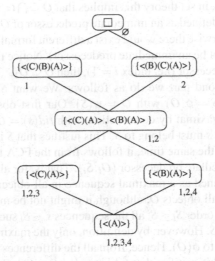

Fig. 3.5 A concept lattice where one of its nodes does not have an initiating stable sequence

after this, S is completed with the maximal sequences of immediate predecessors. But as mentioned above, there might be nodes in the lattice where $\tilde{S} = \emptyset$, so that S cannot be initiated for such construction. In other words, there may be concepts (O,S) with no stable sequence having tid list exactly O. Then, S is just the set of maximal elements in the union of the intents of immediate predecessors.

This situation is illustrated in Figure 3.5, where we present a very simple example of sequential data. In the lattice of this figure there is a closed set of sequences, namely $\{\langle (C)(A) \rangle, \langle (B)(A) \rangle\}$, formed simply as the union of the sequences of its two immediate predecessors. So, we have that neither sequence $\langle (C)(A) \rangle$ nor $\langle (B)(A) \rangle$ has a tid list equal to $\{1,2\}$, which is the closed set of objects linked by the Galois connection. Indeed, the tid list for the stable sequence $\langle (C)(A) \rangle$ is $\{1,2,3\}$, and the tid list for the stable sequence $\langle (B)(A) \rangle$ is $\{1,2,4\}$, but they together only occur in the closed set of objects $\{1,2\}$. In this case, we have a node (O,S) where $tid(s) \subset O$ for each $s \in S$; this ensures a system closed under intersection.

Seq id	Input sequences
d_1	$\langle (C)(B)(C)(A)(C) \rangle$
d_2	$\langle (C)(B)(A)(C)(C)(C)(A) \rangle$
d_3	$\langle (A)(C)(A)(C)(C)(A)(A)(A) \rangle$
d_4	$\langle (C)(A)(C) \rangle$

(a) Collection of data $\mathscr{D}$

Tid list	Stable Sequences
$\{1,2,3,4\}$	$\langle (C)(A)(C) \rangle$
$\{1,2,3\}$	$\langle (C)(C)(A) \rangle$
$\{1,2,3\}$	$\langle (C)(C)(C) \rangle$
$\{2,3\}$	$\langle (A)(C)(C)(C)(A) \rangle$
$\{2,3\}$	$\langle (C)(A)(C)(C)(A) \rangle$
$\{1,2\}$	$\langle (C)(B)(A)(C) \rangle$
$\{1,2\}$	$\langle (C)(B)(C)(A) \rangle$
$\{1,2\}$	$\langle (C)(B)(C)(C) \rangle$
$\{1\}$	$\langle (C)(B)(C)(A)(C) \rangle$
$\{2\}$	$\langle (C)(B)(A)(C)(C)(C)(A) \rangle$
$\{3\}$	$\langle (A)(C)(A)(C)(C)(A)(A)(A) \rangle$

(b) Stable sequences and their tid lists

Fig. 3.6 New example of sequential data with the derived stable sequences

3.5.1 An Algorithm

To better understand Theorem 3.2, here we present the construction of this lattice from a more algorithmic point of view. We are assuming that we are given the set of stable sequences extracted by any of the existing algorithms and we just need to postprocess them. This post-processing will be done by comparing the tid list of the stable sequences and properly organizing them into formal concepts. Since these formal concepts will be constructed incrementally, we propose to rename this term related to the FCA theory as *valid pair*. This redefinition will help to relax the notation, and it will allow the construction of the set of objects as tid lists. Later we will formally prove that, indeed, valid pairs and formal concepts in the lattice are equivalent.

Definition 3.7. A valid pair (T, S) is one where: S is a set of all nonredundant stable sequences whose tid lists are at least T; and $T = \bigcap tid(s), s \in S$.

We say that a set S of stable sequences is *nonredundant* when $S = \max(S)$, that is, all the sequences in S are maximal in the set. By computing the list of valid pairs (T, S) from the set of stable sequences, we get maximal nonredundant groups S associated to a maximal set of transaction identifiers where all sequences in S coexist together. Thus, the idea is that each valid pair will represent eventually a formal concept.

By a way of example, in Figure 3.6 we present a new set of input sequences $\mathscr{D}$, along with the set of all stable sequences and their tid lists. We see that, for example, stable sequences $\langle (A)(C)(C)(C)(A) \rangle$ and $\langle (C)(A)(C)(C)(A) \rangle$, will be fitted into the same set S since they have the same tid list, namely $\{2,3\}$. Any other closed sequence whose tid list is at least $\{2,3\}$ (i.e. coexisting with those sequences) would not fit in S, since it would turn it redundant. Moreover, the set $\{2,3\}$ is a maximal set of transaction identifiers for this S as well, that is, for no other larger set of transactions we have that the sequences in S coexist together. A complete example of

T	S
$\{1,2,3,4\}$	$\{\langle(C)(A)(C)\rangle\}$
$\{1,2,3\}$	$\{\langle(C)(A)(C)\rangle, \langle(C)(C)(A)\rangle, \langle(C)(C)(C)\rangle\}$
$\{2,3\}$	$\{\langle(A)(C)(C)(C)(A)\rangle, \langle(C)(A)(C)(C)(A)\rangle\}$
$\{1,2\}$	$\{\langle(C)(B)(A)(C)\rangle, \langle(C)(B)(C)(A)\rangle, \langle(C)(B)(C)(C)\rangle\}$
$\{1\}$	$\{\langle(C)(B)(C)(A)(C)\rangle\}$
$\{2\}$	$\{\langle(C)(B)(A)(C)(C)(C)(A)\rangle\}$
$\{3\}$	$\{\langle(A)(C)(A)(C)(C)(A)(A)(A)\rangle\}$

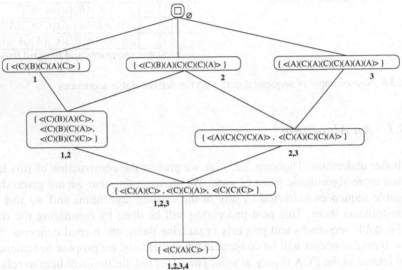

Fig. 3.7 Valid pairs for data in Figure 3.6, and their organization in a lattice

the desired valid pairs for the stable sequences in Figure 3.6(b), is shown in Figure 3.7; these valid pairs exactly coincide with the formal concepts in the lattice that we want. Note that there is no valid pair involving the tid list $\{1,3\}$; this list is not maximal for the set of stable sequences whose tid list is at least $\{1,3\}$. In terms of the closure operator we would say that this list of identifiers does not correspond to a closed set of objects.

Obviously, an initial naive algorithmic approach is to group all those stable sequences with the same tid list, that is, create each pair (T,S) so that each $s \in S$ has the same tid list and $T = tid(s)$. In this way we ensure that each set T is maximal. However, the set S created in the naive way may be incomplete, with some missing stable sequence. In particular, there might be some stable sequences whose tid list is at least T, and that do not make the set S redundant.

Then, a naive grouping such as $S_1 = \{\langle(A)(C)(C)(C)(A)\rangle, \langle(C)(A)(C)(C)(A)\rangle\}$ and $T_1 = \{2,3\}$ of the last example, is a valid pair. However, another naive grouping such as $S_2 = \{\langle(C)(C)(A)\rangle, \langle(C)(C)(C)\rangle\}$ and $T_2 = \{1,2,3\}$, does not form a valid pair: the stable sequence $\langle(C)(A)(C)\rangle$ has tid list $\{1,2,3,4\}$, so it is coexisting with

---

Algorithm 1. Grouping Stable Sequences into Valid Pairs

Input: List CS of stable sequences (extracted by e.g. CloSpan)
Output: List of pairs (T, S)

```
 1:  Sort CS in descending order by tid list;
 2:  while CS ≠ ∅ do
 3:      S ← Next sequences s ∈ CS with same tid list;
 4:      T ← tid(s), for some s ∈ S;
 5:      for each (T', S') immediate predecessor of (T, S) do
 6:          for each s' ∈ S' do
 7:              if s' ⊈ s, for each s ∈ S then S ← S ∪ {s'} end if
 8:          end for
 9:      end for
10:      output (T, S);
11:  end while
```

the other two in the same objects and including it in S_2 does not make it redundant. Because of this, $\langle (C)(A)(C) \rangle$ belongs to S_2 as an element, but not to S_1, where it is already included as a subsequence of $\langle (C)(A)(C)(C)(A) \rangle$.

Of course, here we are finding one of the properties already defined by the theory: there can be stable sequences that belong to several sets S as long as S is not redundant. To identify when a stable sequence must be included in more than one set we will make use of the ordering $\preceq$: given two valid pairs (T', S') and (T, S), if $T \subseteq T'$ then for all $s' \in S'$ there must exist $s \in S$ s.t. $s' \subseteq s$. In other words, we must have $S' \preceq S$. When the valid pairs are proved to be formal concepts this will lead to an organization in the lattice, i.e. (T', S') is a subconcept of (T, S). So, by following this order, just a wise use of lists and indexes turns to be enough to come up with the right groupings.

The pseudocode to obtain valid pairs is presented in Algorithm 1. The idea is simple: lines 3-4 get naive sets of stable sequences with same tid list. Then, lines 5-11 complete the set S with sequences already belonging to immediate predecessors that should also be in S. The stable sequence initializing S and T in lines 3 and 4 serves as our "anchor" to construct the formal concept. Lines 4-11 complete S with the rest of the required elements. Considering that the list of stable sequences, named CS in the pseudocode, is ordered in descending order by the tid list of its elements, the algorithm traverses the conceptual graph bottom-up in a breadth-first fashion. Notice that it is only necessary to look for immediate predecessors of a set S to make the order $\preceq$ hold for all the proper pairs. The complexity is bounded by $O(n \cdot m \cdot k^2)$, where we consider n to be the number of closed sequences in CS, m is the maximum number of immediate predecessors for a node (T, S), and k is the maximum number of closed sequences belonging to immediate predecessors of S.

It is easy to see that Algorithm 1 constructs those valid pairs (T, S) whose set T corresponds exactly to $T = \max\{tid(s) | s \in S\}$. It can be proved that each one of the valid pairs corresponds to closed concepts of the Galois lattice.

Proposition 3.8. *Valid pairs generated by Algorithm 1 are closed concepts.*

Proof. For each valid pair (T, S) we have that $T = \max\{tid(s) | s \in S\}$, and S is a nonredundant set of all stable sequences whose tid is at least T. In other words, if we see the set of transactions T as object identifiers, then this is the same as $\phi(T) = S$. On the other hand, sequences in S are contained simultaneously in transactions T, since $T = \max\{tid(s) | s \in S\}$. Moreover, this set T is maximal by construction, so that $\psi(S) = T$. Thus, (T, S) is also a formal concept. $\square$

However, the construction of valid pairs as in Algorithm 1 does not necessarily yield a closure system. In particular, the reverse implication of Proposition 3.8 may not hold, i.e., there may be closed concepts (O, S) defined by the theory that are not represented by any of the valid pairs derived by the algorithm. This corresponds to the particular case of those formal concepts (O, S) that, as marked by Theorem 3.2, do not have any stable sequence with tid lists exactly O; that is, the set $\tilde{S}$ of the theorem is empty. For example, for the concept lattice of Figure 3.5, Algorithm 1 would not generate the closed set of sequences $\{\langle (C)(A) \rangle, \langle (B)(A) \rangle\}$ since individually, their tid list do not coincide with the closed set of objects linked by the Galois connection.

Therefore, once we get the valid pairs from the algorithm, we may still need to update the system with new concepts that ensure the closure system. This can be done by organizing the set of valid pairs into a temporary lattice, and then, checking the consistency of the dual system of closed sets of objects. Whenever the intersection of two sets of objects produces another set of objects not represented by a valid pair, it will imply that a new concept should be created there. Then, as formalized by Theorem 3.2, the intent of this new concept must be completed by pulling up the maximal stable sequences from the intents of immediate predecessors.

This simple operation of completing the closure system may require checking an exponential number of intersections in the worst case. This extreme situation depends mainly on the distribution of the original data $\mathcal{D}$. Obviously, this data influences in the number of concepts derived from the ordered context, and thus, in the size of the final lattice. The exponential complexity of this problem is similar to the one considered when updating incrementally the lattice with a new instance with binary relationships (see [35, 63, 108, 128]), since there, also new concepts are added and the final closure system has to be properly updated. Later in this documentation, we focus on the empirical validation of such construction. We will see that, in many practical cases the partially ordered set of concepts obtained after Algorithm 1 serves already as a good exploratory aid for the data. For the theoretical results that we will develop in the next chapters, we require a complete closure system.

Finally, a very simple remark: notice that we can see our concept lattice as a natural way of clustering input objects. Indeed, for each node (O, S) of the lattice, we can consider S as a set of features and O as the corresponding input objects projected in that point.

3.6   The Minimum Support Condition

Whereas the restriction to deal with closed sets reduces substantially the number of final patterns, in many cases we still find ourselves confronting large numbers of stable sequences to be examined. Thus, we may need to impose an additional high-frequency condition, θ, over the stable sequences. In such cases, we do not have all the stable sequences but just a set of those over the minimum support, so that we will be interested in constructing only closed concepts corresponding to those frequent stable sequences, i.e. the nodes in the lattice whose extent exceeds θ.

We define the support of a set of sequences S as the number of objects where S is contained, that is, the number of objects linked by the Galois connection, namely $|\psi(S)|$. Alternatively, it can be also defined as follows.

Definition 3.8. The support of a set of sequences S in $\mathscr{D}$ is the number of $d \in \mathscr{D}$ s.t. $S \preceq \{d\}$. We denote this with $supp(S)$.

When considering this minimum threshold θ, we will be interested in those concepts (O, S) with number of objects O over θ. To start the analysis, let us consider a closed set of sequences S. For each $s \in S$ we have that $\{s\} \preceq S$ by Proposition 3.7, so that $supp(S) \leq supp(s)$ by monotonicity. In case there is some stable sequence $s \in S$ with $supp(s) \leq \theta$, then we will have that $supp(S) \leq \theta$ as well. Rephrased in a different way, for all closed sets of sequences with $supp(S) \geq \theta$, we have that $supp(s) \geq \theta$, $s \in S$. Therefore, from frequent stable sequences we can calculate the formal concepts over the threshold.

However, there is an important remark to be made at this point: by monotonicity we know that $supp(S) \leq supp(s)$ for all $s \in S$, yet, $supp(s) \geq \theta$ does not necessarily imply that $supp(S) \geq \theta$. It might be the case that a set of stable sequences S forms a closed set of sequences linked to a closed set of objects O, where each $s \in S$ individually occurs in a number of objects which is a superset of O. Those are the cases where the concept (O, S) is such that O comes from the intersection of the extents of immediate predecessors, and S from the maximal sequences from the union of the intents of immediate predecessors. In terms of Theorem 3.2, we have that $S = \max\{\bigcup S_i\}$, where $\{S_i\}$ are the intents of immediate predecessors indexed by i.

Therefore, from frequent stable sequences extracted by current algorithms we can certainly construct all the frequent concepts by means of Theorem 3.2, but we should prune some of those concepts originated as a consequence of dealing with a closure system, which may not be over the threshold. This minimum support condition is naturally used to reduce the final number of mined patterns in practice; however, from the theoretical point of view, the model constructed with such restriction is not guaranteed to be a complete closure system. This is not a problem if we just want to use the lattice as a visualization tool of the properties of the data; however, for some of our theoretical contributions, we need to ensure formally the properties of the complete closure system. Thus, when considering the analysis of our system, to be done in next chapters, we will be assuming that no minimum support condition has been imposed in the construction of the Galois lattice.

Chapter 4
Horn Axiomatizations for Sequences

This chapter focuses on the study of association rules for ordered data by using the system of closed sets of sequences, defined by operator Δ. Our contribution is a notion of deterministic association rules with order where a set of sequences always implies another sequence in the data. The central advantage of dealing with deterministic rules is that they do not require to select, with little or no formal guidance, one single measure of strength of implication because they always hold. Moreover, since they are pure standard implications, they can be studied in purely logical terms. Indeed, in the second chapter we already mentioned that the set of deterministic association rules derived from classical lattice-theoretic methods axiomatize the minimal Horn upper bound of a binary relation [14]. On the basis of this formalization, the main result of this chapter is a similar characterization of the implications with order as the empirical Horn approximation of the input set of sequences. To allow for this characterization, we will require the definition of certain background Horn conditions to ensure the consistency of the theory. As a consequence of this main result, we can also prove the isomorphy of the lattice of closed sets of sequences and the classical binary lattice when the background Horn conditions hold. Finally, we discuss the computation of all these rules in practice.

4.1 Generators of a Closed Set of Sequences

As in the binary context, a way to generate implications from the lattice is by defining generators. From these generators we will identify the notion of deterministic association rules with order.

Definition 4.1. We say that a set of sequences G is a **generator** of S if we have that $\Delta(G) = S$. We say that a generator G is **minimal** if there is no other G' s.t. $G' \preceq G$ and $G \neq G'$, such that $\Delta(G') = S$.

In our analysis, we will only consider minimal generators, since they do not contain redundancies. These will be graphically added to the concept lattice model with

G.C. Garriga: *Formal Methods for Mining Structured Objects*, SCI 475, pp. 39–50.
DOI: 10.1007/978-3-642-36681-9_4   © Springer-Verlag Berlin Heidelberg 2013

dashed lines. By way of an example, in Figure 4.1 we present a toy example where no repeated items are considered in the input sequences. Minimal generators of the top of the lattice are not shown in the diagram, but, for the sake of illustration, it is easily seen that $\{\langle (C)(B)\rangle\}$ is among them.

Fig. 4.1 A simple example
of ordered data $\mathcal{D}$

Seq id	Sequence
d_1	$\langle (A)(B)(C)(D)\rangle$
d_2	$\langle (B)(C)(D)(A)\rangle$
d_3	$\langle (B)(C)(A)(D)\rangle$

This simple data in Figure 4.1 will ease the follow up of the example, but of course, the theory developed here applies to the general model of data as well. At the end of this chapter we will provide a more complex example to better illustrate our results.

The following lemmas characterize exactly the relation between the generators and their associated closed set of sequences.

Lemma 4.1. *Let $\Delta(G) = S$; then $G \preceq S$ and, for all closed sets of sequences S' s.t. $S' \preceq S$ and $S' \neq S$, we have that $G \npreceq S'$.*

Proof. That $G \preceq \Delta(G)$ follows from the fact that Δ is a closure operator. We prove the following contrapositive of the rest: for closed sets S and S', if $\Delta(G) = S$ and $G \preceq S' \preceq S$ then $S' = S$. Indeed, by monotonicity of Δ, $\Delta(G) \preceq \Delta(S') \preceq \Delta(S)$ and, being S and S' closed, this translates into $S \preceq S' \preceq S$. Using here the fact that all sequences in all closed sets are maximal in them, it follows that $S = S'$. $\qquad\square$

Actually, this is just a rephrasing of the well-known fact that closure operators assign to each set the *minimal* closed set that is above it; in the standard case (unordered

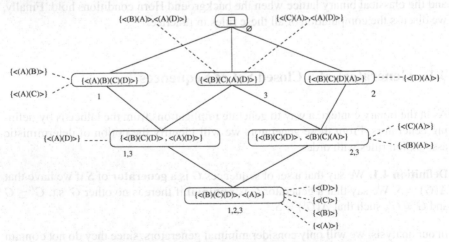

Fig. 4.2 Concept lattice for data in Figure 4.1 with all the minimal generators

data) the comparison is by the standard inclusion of sets, but here the peculiarity is that the comparison is according to $G \preceq S$, as defined in the last chapter.

Lemma 4.2. *Let $G \preceq S$ where S is a closed set of sequences, and assume that, for all closed S', if $S' \preceq S$ and $S' \neq S$ then $G \npreceq S'$; then G contains at least one minimal generator of S.*

Proof. Consider all subsets of G for which the same property indicated for G still holds. Since they are a finite family, at least one of them is minimal in the family (according to $\preceq$). Let G_{min} be this *minimal* subset of G that fulfills the property (or, any of them if there are several): $G_{min} \preceq G \preceq S$, and for all closed $S' \preceq S$ s.t. $S' \neq S$, we have $G_{min} \npreceq S'$. Then, the minimal closed set of sequences containing G_{min} is S, and so, $\Delta(G_{min}) = S$, G_{min} being one minimal generator contained in G.   □

4.2 Empirical Horn Approximations for Ordered Contexts

With the notion of generators set in place we are ready to define our family of deterministic association rules for sequences. We want to prove that this family of rules can be seen also as logical implications axiomatizing the empirical Horn approximation of an enriched theory.

Definition 4.2. A **deterministic association rule with order** is a pair (G, S), denoted $G \to S$ where $G, S \subseteq \mathscr{S}$, such that $\Delta(G) = S$.

We say that a rule $G \to S$ *holds* for a given set of sequences $S' \subseteq \mathscr{S}$ if for all $s' \in S'$ either $G \npreceq \{s'\}$ or $S \preceq \{s'\}$. Due to the construction of the closure operator Δ, we can argue directly that this method of constructing rules is sound, that is, all the rules of our proposed form that can be derived from an input set of sequences $\mathscr{D}$ do hold for each of those input sequences; we could say that our implications with order have confidence 1 (100%) in our data. Indeed, since $\{d\}$ is closed for each individual input sequence d of our database $\mathscr{D}$, we can consider any generator G and obtain, by monotonicity of Δ, $G \preceq \{d\}$ implies $\Delta(G) \preceq \Delta(\{d\}) = \{d\}$; that is, the implication $G \to \Delta(G)$ holds for $\{d\}$.

Now, we need to come back to the propositional logic framework and Horn theories and introduce background knowledge to define the empirical Horn approximation for ordered contexts. To motivate our choices, let us briefly discuss a feature of the analysis in [14]. Indeed, the first step there is to see each unordered transaction as a propositional model, and this is easy to obtain since, actually, it suffices to see the items as propositional variables. We can see this conceptual renaming as an isomorphism, or, even further, by using as propositional variables the very set of items, the translation is a mere identity function.

But this is no longer the case in our ordered contexts. Taking as propositional variables simply the items would not provide a sufficiently structured translation of our data sequences into propositional models. Thus, our next goal is to propose a more specific mapping that considers the ordered context. The resulting empirical

Horn approximation of the ordered data will allow us to characterize the association rules defined in the previous section. By way of an example, consider Figure 4.1, where the first object consists explicitly of the sequence $\langle (A)(B)(C)(D) \rangle$; however, it also contains implicitly all the subsequences $s' \subseteq \langle (A)(B)(C)(D) \rangle$. Thus, each input sequence can be also seen as a tuple of all those subsequences contained in it. Now we assign *one propositional variable to each subsequence* of each input sequence; and restrict the family of possible models by this background knowledge, thus discarding all models that would pretend to include a given sequence s but simultaneously avoid some subsequence of s.

More precisely, let m be a model: we impose on it the constraints that if $m(x) = 1$ for a propositional variable x, then $m(y) = 1$ for all those variables y such that y represents a subsequence of the sequence represented by x. For instance, if a propositional variable x corresponds to the sequence $\langle (A)(B)(C) \rangle$, then a model m assigning 1 to x should also assign 1 to the variable representing $\langle (A)(B) \rangle$, and similarly with other subsequences.

We define more specifically the interpretation of variables as sequences by an *injective* function $\xi : \mathscr{S} \to \mathscr{V}$, where $\mathscr{V}$ is the universe of all variables. For our convenience, we notationally extend this function with $\xi^{-1}(\square) = \square$, where $\square$ is the unsatisfiable boolean constant, corresponding also to the top of our lattice of closed sets of sequences. Now, each input sequence $d \in \mathscr{D}$ in the data corresponds to a model m_d: the one that sets to true exactly the variables $\xi(s')$ where $s' \subseteq d$; and we can find the empirical Horn approximation of the corresponding theory.

Definition 4.3. The set of models from $\mathscr{D}$ is $\mathrm{models}(\mathscr{D}) = \{m_d \mid d \in \mathscr{D}\}$.

It is important that the constraints we have imposed on the models, that when $s' \subseteq s$ then $\xi(s) \to \xi(s')$, are indeed Horn clauses, which we call *background Horn conditions*, and hold on all input models, so that they are imposed automatically on the whole Horn approximation: the conjunction of all Horn clauses satisfied by all the models corresponding to input sequences. We call this conjunction the *empirical Horn approximation for ordered data*, and any model there can be mapped back into a set of sequences that is closed downwards under the subsequence relation.

4.2.1 Main Result

We are now ready to present the equivalence between the association rules extracted by the closure-based method, as presented above, and the empirical Horn approximation for ordered data.

Theorem 4.1. *Given a set of input sequences $\mathscr{D}$, the conjunction of all the deterministic association rules with order constructed by the closure system, seen as propositional formulas, and together with the background Horn conditions, axiomatizes exactly the empirical Horn approximation of the theory containing the set of models $M = \mathrm{models}(\mathscr{D})$.*

Proof. We prove separately both directions for this theorem: 1/ that the deterministic association rules (that is, their corresponding propositional implications) are implied by the empirical Horn approximation; and 2/ that all the clauses in the empirical Horn approximation are implied by the conjunction of the (propositional implications corresponding to) deterministic association rules.

$\Rightarrow$/ Consider a deterministic association rule $G \to S$ s.t. $\Delta(G) = S$. By distributivity, we can rewrite the rule as a conjunction of different implications $G \to s_i$ where $S = \{s_1, \ldots, s_m\} \in 2^{\mathscr{S}}$. As explained above, all the input sequences having as subsequences all the elements of G must have also s_i, so that the translation of $G \to s_i$ is a Horn clause that is true for all the given models in M and, by Theorem 2.1 in chapter 2, it belongs to the empirical Horn approximation. Likewise, the background Horn conditions are also satisfied by all models and thus hold in the empirical Horn approximation.

$\Leftarrow$/ Let $F \to v$ be an arbitrary Horn clause where F is a set of variables, and v is a single variable. Assume this clause to be true for all the given models $M = \{m_d | d \in \mathscr{D}\}$ that correspond to the input sequences; note that these follow the constraints mentioned above: if $m \in M$, and $m(x) = 1$ for a propositional variable x, then $m(y) = 1$ for all those variables y such that $\xi^{-1}(y) \subseteq \xi^{-1}(x)$. In order to show that $F \to v$ is a consequence of the rules found from the concept lattice for $\mathscr{D}$, we will find an association rule that, upon translation, and in the presence of the background Horn conditions, logically implies our Horn clause.

Looking at F as a set of variables, we can consider the set of corresponding sequences $S' = \{\xi^{-1}(v) | v \in F\}$; let $\Delta(S') = S''$ be its closure. By previous lemmas 4.1 and 4.2, we know that S' will contain at least one minimal generator of S'', that is, $G \preceq S'$ s.t. $\Delta(G) = S''$. Therefore, the rule $G \to S''$ will be one of the rules constructed by Definition 4.2. On the other hand, we have assumed that the clause $F \to v$ holds for all the models M. By definition, this means that $S' \to \xi^{-1}(v)$ also holds in all the input sequences, in the sense that whenever $S' \preceq \{d\}$ for an input sequence d, also $\xi^{-1}(v) \subseteq d$; and this implies that $\{\xi^{-1}(v)\} \preceq \Delta(S') = S''$: so, for some sequence $s \in S''$ we have that $\xi^{-1}(v) \subseteq s$ or, equivalently, the Horn clause $\xi(s) \to v$ belongs to the background Horn conditions.

Finally, we have found that $G \to s$ is one of the rules composing $G \to S''$, which is one of the association rules coming from the closure system. Since $G \preceq S'$, the variables corresponding to sequences from G are all in F, and thus the clause $F' \to \xi(s)$ with $F' \subseteq F$ corresponds to one of the association rules. By subsumption, and one resolution step with $\xi(s) \to v$, we see that $F \to v$ follows indeed from the association rules plus the background Horn conditions. $\qquad\Box$

Note that this proof works also well when the Horn clause is nondefinite, that is, when considering $F \to \Box$. In this case, no model from M satisfies all the variables in F, so, $S' \not\preceq \{d\}$ for all $d \in \mathscr{D}$; indeed we have that $\Delta(S') = \Box$ (the top of the lattice not included in any input sequence).

This characterization of Theorem 4.1 brings an immediate consequence: the closure operator of sets of sequences, named Δ, is equivalent to the closure operator for sets of items, named Γ in chapter 2, when the background Horn conditions hold in the considered models. In this case, both lattices turn out to be isomorphic.

Corollary 4.1. *Given a set of input sequences $\mathscr{D}$, let S be a set of sequences and Z be a set of propositional variables such that $Z = \{\xi(s')|s' \subseteq s, s \in S\}$, then, $\Delta(S) = S$ if and only if $\Gamma(Z) = Z$ for the set of models $M = models(\mathscr{D})$.*

Proof. For $\Delta(S) = S$, let $M' = \{m_d|d \in \mathscr{D}, S \preceq \{d\}\} \subseteq M$. By construction, all the variables in Z will be true in each one of the models in M' that keep the background Horn conditions. Moreover, Z is necessarily the intersection of models in M' (because Z is constructed out of S, which is the intersection of input sequences in $\mathscr{D}$ that identify M'); thus we have $\beta(M') = Z$. Also, by construction of Z, this is a set of variables not satisfiable in any other model apart from M', so that $\alpha(Z) = M'$. In case of S being the top of the lattice, by definition we have that $\xi(\square) = \square$, thus getting the unsatisfiable boolean constant representing the top of the binary lattice.

Reciprocally, to prove the other direction of the corollary, let $\Gamma(Z) = Z$, that is, the set Z is a maximal set of variables true in a subset of models named M', where $M' = \{m|m(x) = 1, m \in M, x \in Z\} \subseteq M$. If we consider that these models satisfy the background Horn conditions, we can find the set of input sequences $D' \subseteq \mathscr{D}$ equivalent to M'. For this set we have that $S \preceq \{d\}$ for all $d \in D'$. Again, by construction we necessarily have that S is not included in any other input sequence and it corresponds to the intersection of input sequences in D'. Thus, $\Delta(S) = S$. In the case of dealing with the top element, i.e. $Z = \square$, we always have $\xi^{-1}(\square) = \square$, hence, getting also the top of the lattice for the ordered context. $\qquad\square$

4.3 Computing the Rules with Order

The next step is to discuss the algorithmic solutions for calculating all our implication rules with order. As proved before, the closure operator Δ characterizes the closed patterns of CloSpan (which are closed in the sense of not being extendable in support, thus stable) as those that belong to a closed set. This fact makes CloSpan a good candidate algorithm to construct the concepts of our lattice model. However, computing the deterministic association rules in the ordered data (equivalently, the empirical Horn approximation for the ordered context) we seem to need as well all the minimal generators, in order to output all rules $G \rightarrow S$ where S is closed and G is a minimal generator of S.

Thus, an important step is to add to any current algorithm of mining closed sequential patterns the calculation of minimal generators of each closed set. We want to compute them by means of a general method, so that it can be plugged into any underlying algorithm of mining closed sequential patterns such as CloSpan (or BIDE or TSP). In this way, after computing the closed sets of sequences, the chosen algorithm can directly calculate the minimal generators as well, without incurring inconvenient overheads for intersecting sequences of the database. In this section we show how to compute generators of S as a sort of transversal of appropriately defined differences between S and all immediate closed predecessors in the lattice.

The difficulty of this proposal will depend on the formalization of both steps: first, what is exactly the difference between two sets of sequences, and second, how to properly define the appropriate variant of transversal. The motivation to look for such an approach is that the concept lattice we have obtained is isomorphic to a standard concept lattice (Corollary 4.1) for which such a method of computing rules does already exist [108]; note however that it is not immediate to carry over the isomorphism into the generators, so that we prefer to develop our method fully within the closure operator on sets of sequences.

For comparison purposes, we quote a result that we found in [108] and that we would like to export here, whereby the minimal generators of a closed set in the unordered context obtained by a closure operator Γ are characterized (the original statement differs from ours but their equivalence is readily seen).

Theorem 4.2. *Let Z be a closed set of items $Z = \Gamma(Z)$; the minimal generators of Z are found as the minimal transversal of the hypergraph of the differences $Z - Z'$ where Z' are the immediate closed subsets of Z in the unordered lattice.*

Along this line, other interesting works also developing their results on hypergraph transversals are e.g. [47], [66] or [81]. The transversal hypergraph consists of sets that intersect each and every of the given differences (called *faces* in [108], a term that comes from related matroid-theoretic facts). Also, it is not difficult to see that it suffices to state that the generator intersects the differences with $Z - Z'$ for the immediate closed subsets of Z. For instance, let $Z = ABC$ be a closed set of items, whose immediate closed predecessors in the lattice are $Z'_1 = AB$ and $Z'_2 = AC$; then, the minimal generators of Z can be found by traversing the hypergraph of differences $H = \{Z - Z'_1, Z - Z'_2\}$, that is, $H = \{C, B\}$. The minimal transversal of H is CB, and so it is the minimal generator of Z.

We would like to have a result similar to Theorem 4.2 for the minimal generators of the closed sets of sequences. After Corollary 4.1, we know that a closed set of sequences S can be seen as an equivalent closed set of variables $Z = \{\xi(s')|s' \subseteq s \in S\}$. Therefore, it is possible to characterize the generators of S through the transversals of the hypergraph of differences from the transformed closed Z. For the sake of clarity, here we decide to rewrite this method directly in the language of sequences.

4.3.1 Calculating Minimal Generators for Closed Set of Sequences

We preserve here the term *faces* for our appropriate formalization of the differences between one closed set and its immediate closed predecessors (according to $\preceq$); for closed S, each face of S is $S - S'$, where $S' \preceq S$ is an immediate closed predecessor of S, and the difference is defined as follows:

$$S - S' = \{s | \{s\} \preceq S \text{ but } \{s\} \not\preceq S'\}$$

Note that this operator is similar, but not the same, to the $S \ominus S'$ presented in section 3.5 of the third chapter. There, we defined $S \ominus S' = \{s | s \in S \text{ but } \{s\} \not\preceq S'\}$ to characterize the bottom-up construction of the closed sets of sequences in the lattice. Indeed, it can be readily seen that $S \ominus S' = \max\{S - S'\}$, if S and S' are closed sets of sequences.

The main property now is:

Lemma 4.3. *Let S be a closed set of sequences and $G \preceq S$; then $\Delta(G) = S$ if and only if G intersects all the faces of S.*

Here by G *intersecting a face $S - S'$* we understand set-theoretic intersection, that is, there must exist a common sequence in both. This corresponds to our notion of transversal for ordered data.

Proof. Assume first that G does not intersect the face $S - S'$, for some $S' \preceq S$; thus, no $s \in G$ fulfills the condition in the definition of the face. Since $G \preceq S$, for all such s, $\{s\} \preceq S$ as well, and this implies $\{s\} \preceq S'$, or actually $G \preceq S'$. Now, by monotonicity of Δ, from $G \preceq S' \preceq S$ and the fact that sequences in closed sets are maximal we obtain $S = S'$ just as in Lemma 4.1; and S' is not an immediate predecessor so that $S - S'$ is not a face. Conversely, assume that G indeed intersects all the faces; from $G \preceq S$ and monotonicity again we have $\Delta(G) \preceq S$. Equality will follow as we need, if we prove that $\Delta(G)$ is not an immediate predecessor. Indeed, by Lemma 4.1, $G \preceq \Delta(G)$, so for all $s \in G$, $\{s\} \preceq \Delta(G)$, which negates the condition in the definition of $S - \Delta(G)$. Thus it can't happen that any s is both in G and in $S - \Delta(G)$, and this last difference cannot be a face because G intersects all of them. This implies that $\Delta(G)$ is not an immediate predecessor.   □

Again, we only need to consider immediate predecessors: if G intersects the faces corresponding to immediate predecessors, it must also intersect the other faces, which are larger. Additionally, we may be only interested in minimal generators (according to $\preceq$) since nonminimal generators only yield redundant association rules. However, a result such as Theorem 4.2, that exactly characterizes minimal generators as minimal transversals, does not provide a direct way to compute minimal generators for our closed sets of sequences. In fact, not all the minimal generators that are obtained through Theorem 4.2 in the propositional transformation correspond to minimal generators of the isomorphic closed set of sequences. A clarifying example will be provided in the next subsection.

Note that in order to construct the faces, we only need that, at the time of analyzing a given closed set S, the closed predecessors are known: we do not need the whole lattice. This allows for a sort of incremental processing, whereby as soon as the algorithm has discovered S, it can immediately construct the minimal generators for that S using its immediate predecessors in the lattice. The minimal generators G of a closed S found in this way can be used to construct the association rules of the ordered context, as it was explained above. A more graphical illustration of this construction will be provided in the next subsection.

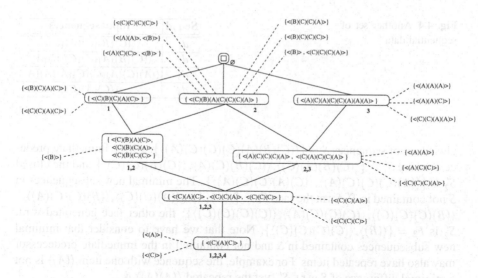

Fig. 4.3 Concept lattice for data in Figure 4.4 with all the minimal generators

4.3.2 Examples

Let $S = \{\langle(B)(C)(A)(D)\rangle\}$ be a closed set of sequences, shown in the lattice of Figure 4.2; the immediate predecessors of S are the closed set of sequences $S_1' = \{\langle(B)(C)(D)\rangle, \langle(A)(D)\rangle\}$ and $S_2' = \{\langle(B)(C)(D)\rangle, \langle(B)(C)(A)\rangle\}$. The minimal new subsequences in S not contained in S_1' are $F_1 = \{\langle(B)(A)\rangle, \langle(C)(A)\rangle\}$, and the minimal new subsequence in S not contained in S_2' is $F_2 = \{\langle(A)(D)\rangle\}$, and these are the two faces. Now, to find the minimal generators of S we must minimally traverse these faces of S, by considering each fragment of order as an atomic variable, and we obtain two generators: $G_1 = \{\langle(A)(D)\rangle, \langle(B)(A)\rangle\}$ and $G_2 = \{\langle(A)(D)\rangle, \langle(C)(A)\rangle\}$, which are exactly the minimal generators of S (see Figure 4.2).

Observe what would have happened if we had applied the method proposed by Theorem 4.2 over the isomorphic propositional transformation of this same node $S = \{\langle(B)(C)(A)(D)\rangle\}$: apart from the two generators G_1 and G_2 mentioned above, we would obtain other sets such as $\{\langle(B)(A)(D)\rangle\}$ or $\{\langle(B)(C)(A)(D)\rangle\}$. These correspond to minimal generators in the transformed space of propositional variables, yet their interpretation as a sequence does not satisfy the minimality condition required to be a minimal generator for the closed set S.

For the sake of illustration, we provide also a more complex example of a set sequences with repeated items. We take the data of Figure 4.4, which was already used in the last chapter to illustrate the construction of the lattice.

The lattice of formal concepts along with the sets of generators are shown in Figure 4.3. For a graphical follow up of the construction of our generators, take the

Fig. 4.4 Another set of
sequential data

Seq id	Input sequences
d_1	$\langle (C)(B)(C)(A)(C) \rangle$
d_2	$\langle (C)(B)(A)(C)(C)(C)(A) \rangle$
d_3	$\langle (A)(C)(A)(C)(C)(A)(A)(A) \rangle$
d_4	$\langle (C)(A)(C) \rangle$

closed set of sequences $S = \{\langle (C)(B)(A)(C)(C)(C)(A) \rangle\}$, whose immediate predecessors are $S'_1 = \{\langle (C)(B)(A)(C) \rangle, \langle (C)(B)(C)(A), \langle (C)(B)(C)(C) \rangle\}$ and the closed
$S'_2 = \{\langle (A)(C)(C)(C)(A) \rangle, \langle (C)(A)(C)(C)(A) \rangle\}$. The minimal new subsequences in
S not contained in S'_1 form the face $F_1 = \{\langle (A)(A) \rangle, \langle (A)(C)(C) \rangle, \langle (B)(C)(C)(A) \rangle,$
$\langle (B)(C)(C)(C) \rangle, \langle (C)(C)(C)(A) \rangle, \langle (C)(C)(C)(C) \rangle\}$; the other face generated w.r.t.
S'_2 is $F_2 = \{\langle (B) \rangle, \langle (C)(C)(C)(C) \rangle\}$. Note that we have to consider that minimal
new subsequences contained in S and not contained in the immediate predecessor
may also have repeated items. For example, the sequence with one item $\langle (A) \rangle$ is not
a minimal difference of S w.r.t. S'_1, yet the repeated $\langle (A)(A) \rangle$ is.

Now, to find the minimal generators of S, the algorithm minimally traverses these
faces of S, by considering each sequence as an atomic variable (that we can identify,
in fact, with $\xi(s)$). The fact of having repeated items does not change the procedure:
the set $\{\langle (C)(C)(C)(C) \rangle\}$ is a minimal generator by itself since it traverses both
faces atomically; on the other hand, we also have sets of more than one sequence,
such as $\{\langle (B) \rangle, \langle (A)(C)(C) \rangle\}$, being also a minimal traversal of the two faces, thus a
minimal generator of S. At this point, we would like to make a special remark about
two minimal transversals: $\{\langle (B) \rangle, \langle (B)(C)(C)(A) \rangle\}$ and $\{\langle (B) \rangle, \langle (B)(C)(C)(C) \rangle\}$.
These generators also atomically intersect both F_1 and F_2, but indeed, they are equiv-
alent to the nonredundant sets $\{\langle (B)(C)(C)(A) \rangle\}$ and $\{\langle (B)(C)(C)(C) \rangle\}$, which are
the ones depicted in the lattice of Figure 4.3. The rest of the generators obtained by
this algorithmic procedure can be followed from there.

Once generators are computed, we will be able to output the set of all deterministic association rules, forming the empirical Horn approximation of the sequential
data. As proposed here, each deterministic rule is created from $G \rightarrow S$ where $\Delta(G) =$
S. For example, let us take the lattice in Figure 4.3. From the bottom node we would
generate two deterministic rules: $\langle (A) \rangle \rightarrow \langle (C)(A)(C) \rangle$ with one of the generators,
and $\langle (C) \rangle \rightarrow \langle (C)(A)(C) \rangle$ with the other generator. Another example is a node of the
same lattice, corresponding to $\{\langle (C)(B)(A)(C) \rangle, \langle (C)(B)(C)(A), \langle (C)(B)(C)(C) \rangle\}$,
and whose unique generator is $\{\langle (B) \rangle\}$. In this case, we also output three deterministic rules corresponding to the generator implying each one of the three sta-
ble sequences in the closed set, that is, $\langle (B) \rangle \rightarrow \langle (C)(B)(A)(C) \rangle$, also $\langle (B) \rangle \rightarrow$
$\langle (C)(B)(C)(A) \rangle$, and $\langle (B) \rangle \rightarrow \langle (C)(B)(C)(C) \rangle$.

Finally, note that because of the construction of such deterministic rules, it might
well be that sequences in the antecedent are not contained in the consequent. For-
mally, for $G \rightarrow s$, where $s \in S$ s.t. $\Delta(G) = S$, there may exist $g \in G$ s.t. $g \nsubseteq s$. For

example, this is the case of a rule such as $\langle (C)(C)(C) \rangle \rightarrow \langle (C)(A)(C) \rangle$, also generated from one of the nodes of the lattice in Figure 4.3.

4.3.3 Reconstructing the Rules of a Cellular Automaton

To explore the capabilities of these closure-based rules, we applied our method to the problem of identifying the rules conforming the local map of a cellular automaton. Broadly speaking, cellular automata are computational systems operating on a space divided into cells, organized into a regular structure (usually, low-dimensional rectangular or hexagonal grids or tori) with a clear, uniform notion of neighborhood of each cell. Cells may change state among a (usually small) number of states, according to the so-called *local map*, the set of rules that govern the evolution of the system along discrete time steps. Each rule in the local map specifies the change of state (or absence of change) of a cell on the basis of the configuration of states of cells in the neighborhood of the cell itself.

We focus our attention on the problem of learning this set of hidden rules that run the evolution of a cellular automaton. Starting from the sequence of evolutions through time and considering that the set of rules that generated such evolution is not known, we would like to discover an approximation of those rules. There are some previous works along this line of analysis. Several of them attempt at modeling textures generated by a sweeping 1D automaton, by identifying "coherent structures" along the spatiotemporal distributions provided by the evolving system, constructing filtering systems for detecting specific phenomena in these evolutions (see [114] and the references there).

Here, we limit ourselves to two-state 1D automata (so, each cell can take one of the two possible values: True or False) where the state of each cell depends only on the states of the two neighbors in the previous generation. We also assume the initial generation to be a random initialization. Each step is encoded by the previous and current states of the left neighbor (l or L), of the right neighbor (r or R) and of the cell itself (c or C), so that each piece of data that the algorithms receive have a form similar to [(l,c,R),(l,C,r)], meaning that at some evolution of the system, at one particular spot, the cell goes from state False (c) to True (C), and at the same time the left neighbor was and remains in state False (l), and the right neighbor changes from True (R) to False (r).

For the sake of illustration, we indicate some of the results on a large automaton working under a disjunctive rule (that is, the state of a cell is the OR of the two states of the neighbors in the previous generation):

- if 'l c' then 'r c', meaning that if the cell is False and in the previous generation its left neighbor is False, then in the previous generation the right neighbor was False as well.
- dually, if 'r c' then 'l c'.

- if 'l C' then 'R C', meaning that if the cell is True and in the previous generation its left neighbor is False, then in the previous generation the right neighbor was True.
- dually, if 'r C' then 'L C'.

Similarly, for the AND function, we get rules such as: if 'L c' then 'r c', meaning that if in the previous generation the left neighbor was True, and the cell is False, then in the previous generation the right neighbor was False as well.

Chapter 5
Transformations on Injective Partial Orders

As mentioned in the first chapter, alternatively to the mining of plain sequential patterns from input sequences, some approaches want to describe portions of the data by means of compatible partial orders, i.e. collections of events occurring frequently together in the input sequences. The complexity of managing these structures and the combinatorial explosion to tackle all the cases makes of this an algorithmically challenging problem.

Here we also address this task by focusing our analysis on the closure system given by operator Δ. As a main result, we will show that each closed set of sequences in the lattice can be derived into the set of maximal paths of a closed partial order. In practice, this transformation yields to an important simplification: algorithms for mining stable sequences, such as CloSpan, can efficiently transform their patterns into closed partial orders, thus avoiding the complexity of the mining operation of those structures directly from the data.

From the theoretical point of view, the transformation of each closed set of sequences into a closed partial order can be formalized by means of operations of category theory [1]. Initially, this chapter will focus on a simplified model where repeated items are not allowed in the set of input sequences; in other words, the partial orders that we are tackling have an injective labelling function and because of this, we name them injective partial orders. In this case, the formalization of the partial order construction from closed sets of sequences simply follows from a coproduct operation of category theory. The simplification of the model made here is meant to ease the presentation and understanding of our results. As we will see in the next chapter, the general case that accepts repeated labels turns out to be far more complicated.

5.1 The Language of Categories

Formally, we will model our partial orders as a full subcategory of the set of directed graphs. As a starting point we recall the most basic among the numerous definitions of graphs given in the literature.

G.C. Garriga: *Formal Methods for Mining Structured Objects*, SCI 475, pp. 51–63.
DOI: 10.1007/978-3-642-36681-9_5   © Springer-Verlag Berlin Heidelberg 2013

Definition 5.1. A **directed graph** is modeled as a triple $G = (V, E, \lambda)$ where V is the set of vertices; $E \subseteq V \times V$ is the set of edges; and λ is the injective labelling function mapping each vertex to an item, i.e. $\lambda : V \to \mathscr{I}$.

The set $\mathscr{I}$ of the labelling function is exactly the finite set of items defined in the introduction; this will be a fixed set in all the graphs belonging to our category. As mentioned above, this chapter considers that the labelling function of a graph is always *injective* (and not necessarily surjective). For the moment, to ease the definitions in this section, we will drop the possibility of having different simultaneous items, that is, we will consider that the labelling function maps each vertex to a single item, not a set of items. For coherence, this corresponds to dealing with single-item sequences. This restriction will be lifted towards the end of this chapter.

A node from a graph $G = (V, E, \lambda)$ will be denoted $u \in V$, or to simplify formalisms we will directly write $u \in G$. An edge between two vertices u and v will be denoted by $(u, v) \in E$ (or for short, $(u, v) \in G$) implying a direction on the edge from vertex u to v; then we also say that node u precedes node v, denoted by $u < v$. As in any category we need morphisms between the considered objects, that is our graphs.

Definition 5.2. A **graph morphism** $h : G \mapsto G'$ between two graphs $G = (V, E, \lambda)$ and $G' = (V', E', \lambda')$ consists of a function $h_V : V \to V'$ that preserves labels (that is, $\lambda = \lambda' \circ h_V$), and $(u, v) \in E \Rightarrow (h(u), h(v)) \in E'$.

All morphisms considered in this chapter are *injective* (that is, function h_V is always *injective*). Note that the injectivity of the morphism between any two graphs, whose labelling function must be also injective, forces the injective morphism h to be unique. So, if there are $h : G \mapsto G'$ and $g : G' \mapsto G$, it implies $h = g$ and $G = G'$. The composition of a morphism $h : G \mapsto G'$ with a morphism $g : G' \mapsto G''$ is the morphism $g \circ h : G \mapsto G''$ consisting of the composed function $g_V \circ h_V$. It is well known that the good properties of graph morphisms turn the set of graphs into a category. From the category of the set of graphs, we will be specially interested here in the following constructor operator.

Definition 5.3 (Coproduct). The coproduct of a family of graphs $\{G_i\}$ indexed by i, is a graph $\widehat{G} = \coprod G_i$ in the same category and a set of morphisms $\{h_i : G_i \mapsto \widehat{G}\}$ such that, for every graph G' and every family of morphisms $\{h'_i : G_i \mapsto G'\}$, there is an unique morphism $g : \widehat{G} \mapsto G'$ such that $g \circ h_i = h'_i$ (see Figure 5.1).

Usually, the index i of the coproduct operation is clear from the context and we remove it to simplify notation. So, when we write $\widehat{G} = \coprod G_i$ we actually mean
$$\widehat{G} = \coprod_{i=1}^{n} G_i$$

In category-theoretic terms, the result $\widehat{G}$ of a coproduct is the *initial object* among all those candidates G' with morphisms $\{h'_i : G_i \mapsto G'\}$. Moreover, we know that this

Fig. 5.1 Coproduct diagram

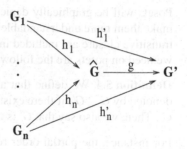

Fig. 5.2 Example of a partial order and its transitive reduction

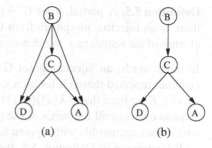

$\hat{G}$ is unique since the family of morphisms $\{h_i\}$, $\{h'_i\}$ and g are injective by definition, and the family of graphs considered here have an injective labelling function.

By construction, the coproduct of two graphs G_1 and G_2 in our category defines exactly a union of G_1 and G_2 where vertices in G_1 and G_2 with the same label are glued together, and where the injectivity of morphisms ensures that all edges from both graphs are preserved.

From the set of all directed graphs, we will be interested in the full subcategory that models partial orders. A *partial order* (also called poset) is a directed graph $G = (V, E, \lambda)$ such that the relation on V established by edges in E is reflexive, antisymmetric and transitive. A *total order* is a partial order that satisfies a fourth property: for any $u, v \in V$, either u before v, $u < v$, or v before u, $v < u$. As in the category of graphs, all the partial orders considered in this chapter have an injective labelling function. The *sources* of a poset are those vertices that do not have any predecessor; similarly, the *sinks* are those vertices which do not precede any other vertex in the poset. Note that a poset may have different unconnected components, and so, several source nodes.

The graphical representation of partial orders is particularly useful for displaying results: we will display a poset by using arrows between the connected labelled vertices, and the symbol $\|$ (parallel) to indicate trivial order among the different components of a poset. The *transitive reduction* of $G = (V, E, \lambda)$ is the smallest relation resulting from deleting those edges in E that come from transitivity (see e.g. [95] for an iterative procedure that returns the transitive reduction of a poset).

Posets will be graphically depicted here by means of their transitive reduction to make them more understandable (as in Figure 5.2(b)), but formally, all edges of the transitive closure are included in E (as in Figure 5.2(a)). Some specific definitions we need on posets are the following ones.

Definition 5.4. We define that a poset G is **more general** than another poset G', denoted by $G \trianglelefteq G'$, if there exists an injective morphism from G to G', i.e. $h : G \mapsto G'$. Then, we also say that G' is **more specific** than G.

For instance, the partial order represented in Figure 5.2 is more specific than the trivial order $A \parallel B \parallel C \parallel D$ (parallelization of all items), but more general than (the transitive closure of) the total order $B \rightarrow C \rightarrow D \rightarrow A$.

Definition 5.5. A partial order $G = (V, E, \lambda)$ is **compatible** with a sequence s if there is an injective morphism from G to the total order obtained as the transitive closure of the sequence s. We denote it as $G \trianglelefteq s$.

In other words, an injective poset $G = (V, E, \lambda)$ is compatible with a sequence s (without repeated items) if for all $u \in V$ we have that item $\lambda(u)$ is in s; and, for all $(u, v) \in E$ we have that $\langle (\lambda(u))(\lambda(v)) \rangle \subseteq s$. The trivial order is compatible with any sequence having all the items of the poset; so, there will be at least one poset (the trivial one) compatible with a given set of sequences.

By extension of Definition 5.5, we also say that a partial order G is *compatible with a set of sequences* $S \subseteq \mathscr{S}$ if $G \trianglelefteq s$, for all $s \in S$. In the case that sequences in S do not have any item in common, then we assume the existence of an empty poset compatible with them. The support of a poset in $\mathscr{D}$ is then the number of input sequences that the poset is compatible with, we denote it by $supp(G)$; similarly, the tid list of a partial order, $tid(G)$, is the list of input sequence identifiers that the poset is compatible with.

Definition 5.6. We define a **path from a poset** G as a sequence s such that there is an injective morphism from the total order obtained from the transitive closure of s to G, i.e. $s \trianglelefteq G$.

Since partial orders are labelled injectively here, the notation used for a path of a poset corresponds simply to an ordered list of labels. By construction each sequence of labels will correspond uniquely to a path of an injective partial order. This will not be the case for general partial orders, where a sequence of labels might identify different paths of the same poset. So, in the next chapter we will need to distinguish between the path of a poset and the sequence of labels associated to the path.

Here, sequences $\langle (B)(C)(D) \rangle$, or $\langle (B)(A) \rangle$, or $\langle (B)(C)(A) \rangle$ define paths of the poset shown in Figure 5.2. Then, we define a path to be *maximal* with respect to the inclusion of sequences. E.g., path $\langle (B)(A) \rangle$ is not maximal since it is a subsequence of the path $\langle (B)(C)(A) \rangle$. Note that posets are acyclic, so, the maximal paths in a poset G will always be defined between sources and sinks of G (of course, avoiding the shortcuts of the transitive closure). Note also that since a poset is actually a graph, we are still able to operate coproducts on them, although this does not necessarily imply that the coproduct of two partial orders is another partial order.

5.2 Coproduct Transformations on Maximal Paths

In this section we want to identify *the most specific* partial orders compatible with a nonempty set of sequences S. Formally, a poset G is the most specific for S if it is the final object in the category of all the posets compatible with S, i.e. any other poset G' compatible with S will be such that $G' \trianglelefteq G$.

When dealing with injective posets, we can establish the uniqueness of the most specific partial order. To prove our results we always assume a nonempty set of sequences $S \subseteq \mathscr{S}$.

Proposition 5.1. *Given a set of sequences $S \subseteq \mathscr{S}$, the most specific poset compatible with S is unique.*

Proof. Let $\{G_i\}$ be the family of injective posets compatible with a set of sequences S, indexed by i. As mentioned after Definition 5.5, at least a trivial order containing the shared items in S belongs to this family. The coproduct of posets in $\{G_i\}$, namely $\widehat{G} = \coprod G_i$, satisfies by construction that $G_i \trianglelefteq \widehat{G}$, for each G_i in the family. Also, it is true by hypothesis that $G_i \trianglelefteq s, s \in S$, so that we can resort to the initiality of the coproduct transformation to certify that $\widehat{G} \trianglelefteq s, s \in S$. Moreover, this coproduct construction over a family of injective posets and injective morphisms is unique: let G' be another injective poset s.t. $G_i \trianglelefteq G'$ and G' compatible with S; because G' belongs to the family $\{G_i\}$ we have that $G' \trianglelefteq \widehat{G}$; yet, we also have that $\widehat{G}$ belongs to this family $\{G_i\}$, so that the reverse morphism $\widehat{G} \trianglelefteq G'$ also exists by hypothesis. Then, $\widehat{G} = G'$ in this category of injective posets, and we can conclude that $\widehat{G}$ is unique and the most specific partial order for S. □

We must point out an important remark after this proof: the coproduct is an operator defined on graphs, so, the coproduct of two partial orders may give a graph that is not another partial order. For instance, the coproduct of $G_1 = A \rightarrow B$ and $G_2 = B \rightarrow A$ leads to a graph with a cycle (thus, not antisymmetric). However, in the proof of Proposition 5.1 we are operating the coproduct on a set of posets simultaneously compatible with the same set of sequences S. Given that the sequences in S do not have repeated items by definition, it is not possible to get a cycle from the coproduct of two, or more, different posets compatible with S; in other words, two partial orders whose union leads to a cycle cannot both be compatible with the same set S.

We are ready now to look for the most specific partial orders, but not directly from the data, but describing them through their maximal paths, as follows.

Theorem 5.1. *The set of maximal paths of the most specific partial order compatible with a set of sequences $S \subseteq \mathscr{S}$, with no repeated items, coincides with the intersection of those sequences in S.*

Proof. Let G be the most specific injective partial order compatible with the set of sequences S. If $s' = \langle (I_1) \ldots (I_n) \rangle$ defines a path of G, by definition there exists an injective morphism $s' \trianglelefteq G$. Because G is stated to be compatible with S, we also have $G \trianglelefteq s$ for all $s \in S$. Then, by composition of the morphisms we get $s' \trianglelefteq s$, for all $s \in S$. This means that $s' \subseteq s$, for all $s \in S$, and thus, s' will be a subsequence of some sequence in the intersection of those S.

Now, let $S' = \bigcap s$, $s \in S$. By construction, the total order obtained by the transitive closure of each $s' \in S'$ has an injective morphism to each $s \in S$, i.e. $s' \trianglelefteq s$. This means that each s' alone belongs to the family of injective posets compatible with S. On the other hand, that G is the most specific partial order for S means that this is the final object in the category of posets compatible with S. Then, we necessarily have that each $s' \in S'$ has a morphism to this final object, i.e. we have $s' \trianglelefteq G$, $s' \in S'$. By Definition 5.6, this means that s' is a path of G.

That is, we have that: for all paths t of the most specific poset G, $t \trianglelefteq s'$ for some $s' \in S'$; and for all $s'' \in S'$, there exists a path t in G s.t. $s'' \trianglelefteq t$. In particular, this means that there exists a maximal path t_{max} in G s.t. $s'' \trianglelefteq t_{max}$. So, by composition of morphisms we get $s'' \trianglelefteq t_{max} \trianglelefteq s'$ for some $s' \in S'$, which necessarily implies $t_{max} = s'' = s'$ (because by construction all sequences in S' are maximal in the set). Dually, there exists $s' \in S'$ s.t. $t_{max} \trianglelefteq s' \trianglelefteq t'_{max}$, for some t'_{max} maximal paths of G, which implies that $s' = t_{max} = t'_{max}$ (because by definition, maximal paths are maximal sequences in the set of all paths). Therefore, this proves that intersections on S coincide exactly with the maximal paths of G. $\square$

This statement is true only in case we are dealing with injective partial orders. When allowing for repeated items in the input sequences, the implication from left to right of this theorem is not necessarily true. More details appear in the next chapter.

To illustrate the result of the theorem, consider the set of two input sequences $\{d_2, d_3\}$ from data in Figure 5.3. The most specific partial order compatible with them is the one shown in Figure 5.2. As is stated by the theorem, the maximal paths of this partial order must coincide exactly with the intersection of $d_2 \cap d_3$, that is, the paths are exactly $\langle (B)(C)(D) \rangle$ and $\langle (B)(C)(A) \rangle$.

Next Theorem 5.2 completes this characterization.

Theorem 5.2. *The most specific partial order compatible for a set of sequences $S \subseteq \mathscr{S}$, is the result of the coproduct transformation on the intersection of sequences in S.*

Proof. Let $S' = \bigcap s$, $s \in S$, and we define $\widehat{G} = \coprod s'$, $s' \in S'$. Also, let G be the most specific poset for S. We want to prove that G coincides with $\coprod s'$, $s' \in S$. It follows from the initiality of the coproduct operation that the morphism $\widehat{G} \trianglelefteq G$ exists and is injective. Moreover, after Theorem 5.1 we know that each $s' \in S'$ defines a maximal path of G, and because all posets are labelled injectively here we have that each node $u \in G$ with a unique label l is mapped by exactly one single node $u' \in \widehat{G}$,

Seq id	Sequence
d_1	$\langle (A)(B)(C)(D) \rangle$
d_2	$\langle (B)(C)(D)(A) \rangle$
d_3	$\langle (B)(C)(A)(D) \rangle$

Fig. 5.3 Example of ordered data $\mathscr{D}$

Fig. 5.4 Example of a coproduct

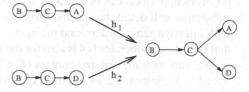

that also has this unique label l. That is, the morphism $\widehat{G} \trianglelefteq G$ is surjective. Thus, $G = \widehat{G} = \coprod s', s' \in S'$.   □

Consider the set of two input sequences $\{d_2, d_3\}$ from data in Figure 5.3, whose intersection corresponds to the set $\{\langle (B)(C)(D) \rangle, \langle (B)(C)(A) \rangle\}$. According to Theorem 5.1, this set coincides with the maximal paths of the most specific partial order compatible with d_2 and d_3 at a time. Then, after Theorem 5.2 we can obtain this most specific poset from the coproduct transformation of these two sequences, which is shown in Figure 5.4. The result exactly corresponds to the poset in Figure 5.2.

5.2.1 The Simultaneity Condition Revisited

To consider input sequences $s = \langle (I_1)(I_2) \dots (I_n) \rangle$ where each I_i may contain several simultaneous items, we must redefine the labelling function λ with $\lambda : V \to 2^{\mathscr{I}}$. Again, this function is required to avoid repetitions of items in the nodes of the partial order (since we are not allowing for repetitions in the input sequences either). So, the injectivity restriction of λ means that each one of the items is unique in the poset. That is, each item $i \in \mathscr{I}$ may be used at most once to label one single node (with or without other simultaneous items). With the new labelling function, the path from a poset is a sequence of itemsets and not single items, and morphisms considered remain still injective. All the results we presented above naturally hold with the new definitions.

5.3 The Closure System of Partial Orders

Next we present the construction and visual display of a concept lattice where nodes contain partial orders, and the relationships among them will be representative of the relationships in the input ordered data. The idea is that each one of the nodes in this lattice corresponds to the formalization of a closed partial order compatible with a maximal set of input sequences.

As customary in FCA, we focus on the characterization of the closure operator specific for partial orders via derivation operators. Notice that this characterization can be formalized without requiring the injectivity constraint needed above, and indeed, the closure operator here is directly characterized for the general model of posets (where labels can be repeated and morphisms may not be injective). The next subsection will detail the specific properties derived from the lattice when dealing with injective partial orders, and the next chapter will provide all the details to state that this closure operator holds also for the general model.

We start with the ordered context $(R, \mathcal{O}, \mathcal{I})$ associated to the data $\mathcal{D}$, as introduced in Definition 3.2 of the third chapter. We recall that objects are denoted as natural numbers, so that $o \in \mathcal{O}$ identifies the input sequence $d_o \in \mathcal{D}$. Let $\mathcal{G}$ identify the universe of partial orders (not necessarily injective).

- For a set $O \subseteq \mathcal{O}$ of objects we define,

$$\sigma(O) = \{G \in \mathcal{G} \,|\, G \text{ is the most specific poset s.t } G \trianglelefteq d_o, \text{ for all } o \in O\}$$

- Correspondingly, for a partial order $G \in \mathcal{G}$ we define,

$$\tau(G) = \{o \in \mathcal{O} \,|\, G \trianglelefteq d_o\}$$

Function $\tau(G)$ returns the set of (indices of) input sequences where G is compatible, that is, we have that $\tau(G) = tid(G)$.

Sets of objects can be partially ordered by the standard inclusion $\subseteq$ (because they simply correspond to indexes), while partial orders will be compared by the morphism-based relation $\trianglelefteq$. Notice that this relation $\trianglelefteq$ defines a partial order only on injective posets; for general posets we simply have a preorder. Now we have the following proposition, corresponding to the properties of Galois connections.

Proposition 5.2. *For sets of objects $O, O' \subseteq \mathcal{O}$, and partial orders $G, G' \in \mathcal{G}$, the following properties hold:*

    1) $O \subseteq O' \Rightarrow \sigma(O') \trianglelefteq \sigma(O)$ 1') $G \trianglelefteq G' \Rightarrow \tau(G') \subseteq \tau(G)$

    2) $O \subseteq \tau(\sigma(O))$ 2') $G \trianglelefteq \sigma(\tau(G))$

Proof. These properties can be reasoned similarly to Proposition 3.2, in chapter 3.

1)   Let $G' = \sigma(O')$. We have that, for all $o' \in O'$, G' is compatible with o', that is, in particular G' is also compatible with o, for all $o \in O$, if $O \subseteq O'$. Thus, there exists $G = \sigma(O)$ s.t. $G' \trianglelefteq G$, which means $\sigma(O') \trianglelefteq \sigma(O)$.

2)   For all $o \in O$ we have that $G \in \mathcal{G}$ is compatible with o, where $G = \sigma(O)$, so that $o \in \tau(\sigma(O))$. Thus, $O \subseteq \tau(\sigma(O))$.

1')   For all $o' \in \tau(G')$ we have that G' is compatible with o', that is, in particular we also have that G is compatible with o' if $G \trianglelefteq G'$. Thus, $o' \in \tau(G)$, which means $\tau(G') \subseteq \tau(G)$.

2')   Given $G \in \mathcal{G}$, it is compatible with o for all $o \in \tau(G)$. Then, if $G' = \sigma(\tau(G))$, by construction we have $G \trianglelefteq G'$, which implies $G \trianglelefteq \sigma(\tau(G))$. $\square$

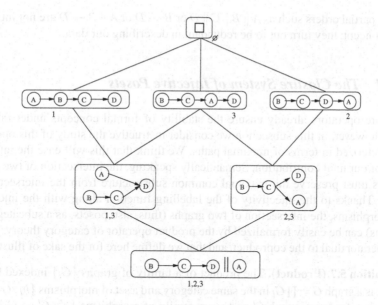

Fig. 5.5 Concept Lattice of partial orders for data in Figure 5.3

From these properties we have two closure systems that are dually isomorphic to each other. Monotony, extensivity and idempotency follow immediately for both compositions of σ and τ after Proposition 5.2, and here we omit this proof for simplicity.

Proposition 5.3. *Compositions* $\widehat{\Omega} = \tau \cdot \sigma$ *and* $\Omega = \sigma \cdot \tau$ *are closure operators.*

Our operator of interest here is Ω, which works on partial orders. As usual, we have that *closed partial orders* are those coinciding with their closure, that is, $\Omega(G) = G$. Formal concepts of the form (O, G) will be nodes of the concept lattice. They correspond to the most specific partial orders associated to a maximal set of objects. We will visualize these nodes by the closed poset G of the concept, and the dual O will be added as a label to the node. Edges in the lattice will be the specificity relationships among the concepts: (O_1, G_1) is a subconcept of (O_2, G_2) if $G_1 \unlhd G_2$ (equivalently $O_2 \subseteq O_1$). As is common in FCA, an artificial top representing a poset not belonging to any sequence is also added to the lattice. Again, we denote it by the unsatisfiable boolean constant $\square$.

In Figure 5.5 we show the lattice of closed concepts for the data of Figure 5.3, where items are not repeated. For this simple case, we see that the concept lattice of partial orders completely summarizes the data. It presents a balance between generality and specificity for all the input sequences. For example, the partial order in the bottom of the lattice is compatible with all the input sequences, but it is less specific than any of the partial orders located above, which are compatible with two data sequences; total orders are compatible with just one single input sequence.

Also, partial orders such as $A \parallel B \parallel C \parallel D$ or $B \rightarrow D$ or $A \parallel B \rightarrow D$ are not intents of any concept; they turn out to be redundant in describing our data.

5.3.1 The Closure System of Injective Posets

Closure operators already ensure the stability of formal concepts under intersection; however, in this subsection we consider instructive the study of this operation characterized in terms of maximal paths. We think that this will ease the introduction of our final contribution. Semantically speaking, the intersection of two partial orders must preserve the maximal common substructure from the intersected objects. Thanks to the injectivity of the labelling function along with the injectivity of morphisms, the intersection of two graphs (thus, also posets, as a subcategory of graphs) can be easily formalized by the product operator of category theory. This is an operator dual to the coproduct, and that we define here for the sake of illustration.

Definition 5.7 (Product). The product of a family of graphs $\{G_i\}$ indexed by $1 \leq i \leq n$, is a graph $\widehat{G} = \prod G_i$ in the same category and a set of morphisms $\{h_i : \widehat{G} \mapsto G_i\}$ such that, for every graph G' and every family of morphisms $\{h'_i : G' \mapsto G_i\}$, there is an unique morphism $g : G' \mapsto \widehat{G}$ such that $h_i \circ g = h'_i$ (see Figure 5.6).

Note that the diagram of a product just reverses the morphisms of the coproduct operator. Also for the product, we know that the resulting $\widehat{G}$ is unique since the family of morphisms $\{h_i\}$, $\{h'_i\}$ and g are injective, and the family of graphs considered here have an injective labelling function. Therefore, the product of two posets G and G' defines exactly their intersection. An alternative and particularly instructive way, as we shall see, is to present the intersection of injective partial orders in terms of the coproduct of their maximal paths.

Definition 5.8. We define the intersection of two injective partial orders $G \cap G'$ as $\coprod s'_i$, where $\{s'_i\}$ is the family of maximal paths common to both G and G' (i.e., $\{s'_i\}$ contains those maximal sequences from the family of all common paths to G and G').

The next lemma proves that the intersection as it is defined here preserves the maximal common substructure from the intersected partial orders, that is, it is equivalent to a product on the partial orders.

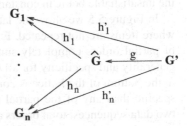

Fig. 5.6 Product diagram

Lemma 5.1. *The result of $G \cap G'$ defined as the coproduct of the family of maximal paths common to them, corresponds to $G \prod G'$.*

Proof. Let $\widehat{G} = \coprod s'_i$, where $\{s'_i\}$ is the family of maximal paths contained in both G and G' (Definition 5.8). By hypothesis, injective morphisms $s'_i \trianglelefteq G$ and $s'_i \trianglelefteq G'$ exist for each s'_i in the defined family of maximal common paths; thus, there exist also injective morphisms $s'_i \trianglelefteq G \prod G'$ for each s'_i. Because the coproduct returns an initial object, we have that $\widehat{G} \trianglelefteq G \prod G'$ exists and it is injective. Moreover, all posets are labelled injectively here and we can resort to the maximality of each path s'_i in the family to state that this morphism $\widehat{G} \trianglelefteq G \prod G'$ is necessarily surjective (i.e., each node $u \in G'$ with unique label l is mapped exactly by one single node $u' \in \widehat{G}$, that also has unique label l). Therefore, $G \prod G' = \widehat{G}$. $\square$

It follows from closure operator, this lattice is closed under intersection.

Proposition 5.4. *The intersection of injective closed posets is another injective closed poset.*

Proof. Let G and G' be two different closed partial orders, and let $\{G_i\}$ be the family of closed posets in the lattice such that $G_i \trianglelefteq G$ and $G_i \trianglelefteq G'$. Now, we define the poset intersection of G and G', i.e. $\widehat{G} = G \prod G'$. By construction of the closure operator we have that $\Omega(\widehat{G}) = G_i$, for some G_i in the defined family. This means that $\widehat{G} \trianglelefteq G_i$. On the other hand, by construction of the product, $\widehat{G}$ is always a final object, so that the reverse morphism must exist too, i.e. $G_i \trianglelefteq \widehat{G}$. Therefore, one of the closed posets in the family is indeed $\widehat{G}$, i.e. there exists $G_i = \widehat{G}$. $\square$

5.4 Lattice Isomorphy

With the lattice of closed partial orders and all its properties set on place, we are ready for the last contribution of this chapter. Naturally, it follows from Theorem 5.2 that each closed set of sequences represents the maximal paths of the closed partial order associated to the set of objects in the same concept. Formally, we prove this as follows.

Theorem 5.3. *The lattice of closed sets of sequences in $\mathscr{D}$ (where no repetitions of items are allowed) and the lattice of closed partial orders in $\mathscr{D}$ are isomorphic.*

Proof. By definition, $\phi(O) = S$ corresponds to the intersection of those input sequences in $\mathscr{D}$ identified by O. It follows immediately after Theorem 5.2 that S can be transformed via coproducts into the *most specific* partial order compatible with those sequences identified by O, namely G. Moreover, this transformation is unique after Proposition 5.1, and along with the definition of σ, we have that $\sigma(O) = G$. Also, by definition of $\psi(S) = O$ the set of objects O is maximal for S, and by definition of τ this implies that $\tau(G) = O$. Thus, (O, G) is a formal concept in the lattice of closed partial orders.

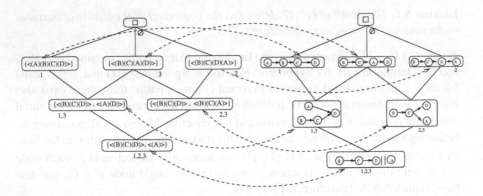

Fig. 5.7 Lattice transformation to closed partial orders

On the other hand, let S be the set of maximal paths of the closed poset G. Given that $\tau(G) = O$, we necessarily have that $\psi(S) = O$. Also, given that $\sigma(O) = G$, this implies by construction that $\phi(O) = S$. Thus, (O, S) is a formal concept in the lattice of stable sequences.   □

Figure 5.7 shows the set of all stable sequences from data in Figure 5.3 organized in formal concepts. Dashed lines of the figure indicate, on one direction, a node rewriting with the corresponding coproduct, and on the other direction, the exact equivalence with the maximal paths. Note that the support of the generated closed partial order will coincide with the support of the closed set of sequences.

The result of Theorem 5.3 means that the set of all injective closed partial orders in $\mathscr{D}$ can be obtained by identifying positions of the sequences in a closed set, as long they share the same label (a label being a single item, or a set of items when considering simultaneity). This will ensure the output of the most specific partial order, and, by the properties studied above, we will not get any cycle. Closed partial orders over a specific minimum support threshold can be obtained form the frequent concepts of the lattice.

Notice that the isomorphy between these two lattices allows for a different interpretation of the intersection of two, or more, closed posets. Indeed, the intersection of a set of closed partial orders can be defined also in the isomorphic universe of closed sets of sequences; this will correspond to intersecting sets of maximal paths and performing a coproduct on them. This idea corresponds closely to Definition 5.8.

5.4.1 Examples

As mentioned early in the chapter, we can also consider simultaneity in the input sequences. All the results defined above hold for the general model without repeated items, but considering simultaneity, i.e. the labels of the closed partial orders may correspond to a set of items. To illustrate the construction of closed partial orders when simultaneity is allowed, let us consider the example in Figure 5.8.

The transformation process between the two lattices is depicted in Figure 5.9. Each stable sequence $\langle (I_1)(I_2)\ldots(I_n)\rangle$ of a closed set of sequences is transformed into a maximal path of a partial order. Since items in the sequences will not be repeated, then, the labelling function of each partial, i.e. $\lambda : V \rightarrow 2^{\mathscr{I}}$ is also injective. In other words, each individual item is part of the label of one single node.

Fig. 5.8 Example of ordered data $\mathscr{D}$ with simultaneity and without repeated items

Seq id	Sequence
d_1	$\langle (CD)(A)(B)\rangle$
d_2	$\langle (A)(CD)(B)\rangle$
d_3	$\langle (D)(C)(A)(B)\rangle$

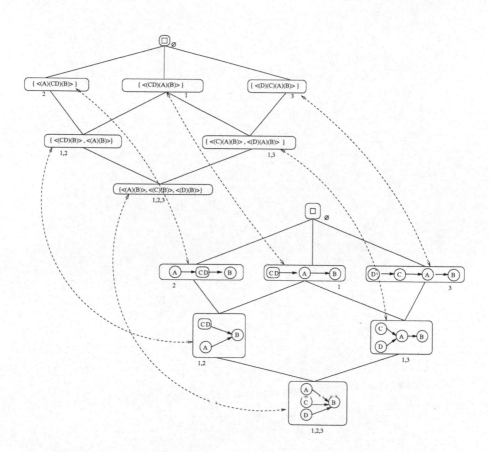

Fig. 5.9 Concept lattice transformation for data in Figure 5.8

corresponds to a set of items. To illustrate the construction of closed partial orders when simultaneity is allowed, let us consider the example in Figure 5.8.

The transformation process between the two lattices is depicted in Figure 5.9. Each stable sequence $\langle Q_1(t_1) \ldots Q_r(t_r) \rangle$ of a closed set of sequences is transformed into a maximal path of a partial order. Since items in the sequences will not be repeated then, the labelling function of each partial label $\{c, A, B\} = \{c, A, B\}$. It also specifies in other words, each individual item is part of the label of one single node.

Seq Id	Sequence
q_1	$\langle (C,D)(A)(B) \rangle$
q_2	$\langle (A)(C,D)(B) \rangle$
q_3	$\langle (D)(C)(A,B) \rangle$

Fig. 5.8 Example of ordered data-set with similar labels and without repeated items

Fig. 5.9 Concept lattice transformation for data in Figure 5.8

Chapter 6
Transformations on General Partial Orders

Next we will be considering repetitions of items in the input sequential data, and as a consequence, the final closed partial orders are not necessarily injective. As we will see, dealing with general partial orders makes the proper formalization with category theory a bit more difficult. To start with, we are forced to drop the injectivity of the morphisms in the general category of graphs, and this allows for many different ways of mapping a partial order over a sequence. Still another inconvenience, Theorem 5.1 of chapter 5 just holds for one of the directions in this new problem. Indeed, the maximal paths of the final closed partial orders do not necessarily coincide exactly with the intersections of the compatible input sequential data. Here we will try to formally justify the construction of our final closed partial orders for a set of data as the colimit transformation on path-preserving edges. Colimits will naturally generalize also the coproduct transformation of chapter 5, but as we shall see, proving that our final structure has the property of being maximally specific is still an unsolved combinatorial problem.

6.1 Notational Review

We will use the same notation and terms already defined in chapter 5, but adapted to the general model where items can be repeated. Since the basic formalizations still hold, in this section we specify which new definitions and differences are considered for the new problem. These differences mainly concern the family of morphisms.

Definition 6.1. A **directed graph** is modeled as a triple $G = (V, E, \lambda)$ where V is the set of vertices; $E \subseteq V \times V$ is the set of edges; and λ is the labelling function mapping each vertex to an item, i.e. $\lambda : V \to \mathscr{I}$.

Partial orders are modeled as a full subcategory of the set of directed graphs, but now the labelling function of the graph is not necessarily injective, that is, there might be several vertices labelled with the same item. For the moment, we consider that our labels are single items, that is, no simultaneity is allowed. As in the last

G.C. Garriga: *Formal Methods for Mining Structured Objects*, SCI 475, pp. 65–83.
DOI: 10.1007/978-3-642-36681-9_6   © Springer-Verlag Berlin Heidelberg 2013

(a) Poset 1   (b) Poset 2

Fig. 6.1 Example of two general partial orders

chapter, this restriction will be lifted later on. An example of two partial orders is depicted in Figure 6.1.

For our convenience, we do not consider self-loops in the definition of a graph, i.e. $u \neq v$ for all $(u,v) \in G$. This choice will be justified below. As in any category we need to define morphisms between the graphs.

Definition 6.2. A **graph morphism** $h : G \mapsto G'$ between two graphs $G = (V,E,\lambda)$ and $G' = (V',E',\lambda')$ consists of an function $h_V : V \to V'$ that preserves labels (that is, $\lambda = \lambda' \circ h_V$), and $(u,v) \in E \Rightarrow (h(u),h(v)) \in E'$.

These morphisms will play an important role to define the specificity and generality of our structures. In the last chapter only injective morphisms sufficed to compare partial orders because each label was unique; however, with general graphs, we will need to break this injectivity to ensure the correctness of the results. We will talk about injective morphisms between graphs when the function h_V of the above definition is injective. If nothing is specified we mean that the morphism may be not injective. Notice that contrary to chapter 5, if we have $h : G \mapsto G'$ and $g : G' \mapsto G$, this does not imply $G = G'$.

Also, it is worth noting that because we are not considering self-loop edges, it is not possible to map $(u,v) \in G$ with $u \neq v$ but $\lambda(u) = \lambda(v)$, into a single node of another graph. This is justified by the choice on the definition of subsequence (discussed already in the first chapter): for $s \subseteq s'$ we do not allow different adjacent positions of s to be included into one single position of s' (this is the definition proposed by Agrawal and Srikant in [5, 117]). When sequences are seen as the transitive closure of total orders, this implies that self-loops are not considered to map s into s'. In other words, we want $s \subseteq s'$ to be equivalent to a morphism $h : s \mapsto s'$.

From the category of the set of graphs and the universe of morphisms, we will be interested here in the following constructor operator.

Definition 6.3 (Colimit). For any diagram containing the family of graphs $\{G_i\}$ and morphisms $\{g_i\}$, the colimit of this diagram is a graph $\widehat{G}$ and the family of morphisms $\{f_i\}$, such that for each $f_i : G_i \mapsto \widehat{G}$, $f_j : G_j \mapsto \widehat{G}$, and $g_k : G_i \mapsto G_j$, we have that: $f_j \circ g_k = f_i$, and $\widehat{G}$ is the initial object in the full subcategory of all such candidates G'.

Since we deal with the general category of graphs, morphisms are not necessarily injective nor surjective in this diagram. We define the following relationship between

Fig. 6.2 Colimit diagram

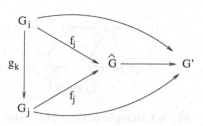

partial orders, that basically, is the same used in chapter 5 but with the general notion of morphism.

Definition 6.4. We define that a poset G is **more general** than another poset G', denoted by $G \trianglelefteq G'$, if there exists a morphism from G to G', i.e. $h : G \mapsto G'$. Then, we also say that G' is **more specific** than G.

This definition makes the poset in Figure 6.1(a) more general than the total order depicted in Figure 6.1(b): two nodes labelled with a C in the first poset collapse in the first C of the second poset. Note that if we had imposed injective morphisms to compare the partial orders, then there would not exist any morphism available between these two posets of the example, and as a consequence, we would not be able to decide which one is the most specific. Also, notice that our setting allows the mapping of two nodes u, u' of G with the same label, into the same node of G' only if they are not comparable in G, due to the absence of self-loops.

Naturally from here, we also define the compatibility of a partial order with a sequence. An obvious observation that we already used in the last chapter is to realize that (the transitive closure of a) sequence s is a total order. Then:

Definition 6.5. A partial order $G = (V, E, \lambda)$ is **compatible** with a sequence s if there is a morphism from G to the total order obtained as the transitive closure of the sequence s. We denote it as $G \trianglelefteq s$.

For example, we have that the partial order in Figure 6.1(a) is compatible with the sequence $\langle (A)(C)(B)(A)(C) \rangle$, or also with the sequence $\langle (C)(B)(A)(C)(A)(C) \rangle$. By extension of Definition 6.5, we also say that a partial order G is *compatible with a set of sequences* $S \subseteq \mathscr{S}$ if $G \trianglelefteq s$, for all $s \in S$. Again, the support of a partial order G in $\mathscr{D}$ corresponds to the number of input sequences where G is compatible; we denote this by $supp(G)$.

Because in this chapter we need to distinguish between the maximal path of a partial order and a sequence, we will redefine the notion of path as a sequence of nodes, not a sequence of labels that was used in last chapter.

Definition 6.6. We define a **path from a poset** $G = (V, E, \lambda)$ as the total order $u_1 \to u_2 \ldots \to u_n$, where $u_i \in V$ and there is $(u_i, u_{i+1}) \in E$.

If t is a path from a poset G we always have that there is an injective morphism $t \trianglelefteq G$. Moreover, a path $u_1 \to u_2 \ldots \to u_n$ from G is *maximal* if u_1 is a source node, and

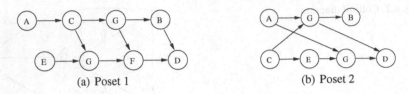

(a) Poset 1   (b) Poset 2

Fig. 6.3 Example of two partial orders: Figure 6.3(a) corresponds to a poset with repeated maximal paths, and Figure 6.3(b) is a poset having a maximal path which is not a maximal sequence

u_n is a sink node and $(u_i, u_{i+1}) \in E$ is an edge belonging to the *transitive reduction* of the partial order G. The transitive reduction of a poset is unique and it can be found iteratively by removing redundant edges coming from the transitive closure, see e.g. [95]. Note that paths are always length bounded since self-loop edges are not considered in our formalization.

We say that the *sequence of labels* associated to a path $u_1 \to u_2 \ldots \to u_n$ from $G = (V, E, \lambda)$ is the sequence $\langle (\lambda(u_1))(\lambda(u_2)) \ldots (\lambda(u_n)) \rangle$. Note that in general partial orders we need to distinguish specifically between a maximal path from its associated sequence of labels. Indeed, a sequence of labels may correspond to different maximal paths in general partial orders. For example, this is the case of poset in Figure 6.3(a), where the sequence $\langle (A)(C)(G)(F)(D) \rangle$ corresponds to two maximal paths of the same partial order. Also, we may have partial orders with maximal paths not corresponding to maximal sequences; e.g. the poset in Figure 6.3(b), where the sequence of labels $\langle (C)(G)(D) \rangle$ corresponds to one of its maximal paths, but it is not a maximal sequence since it is included into another the sequence of labels, namely $\langle (C)(E)(G)(D) \rangle$, corresponding to another maximal path of the same poset.

We also require the following new definitions. The *head* of $s = \langle (I_1) \ldots (I_n) \rangle$ up to a position j s.t. $1 \le j \le n$, is denoted with $head(s, j) = \langle (I_1) \ldots (I_j) \rangle$. Similarly, the *tail* of s from position j s.t. $1 \le j \le n$, is denoted with $tail(s, j) = \langle (I_j) \ldots (I_n) \rangle$. We denote by $s[i]$ the *position i* of s. The *concatenation* of two sequences will be $s \diamond s'$.

6.2 Partial Orders to Summarize Sequences

As in the last chapter the goal is to find a good summary of a nonempty set of sequences $S \subseteq \mathscr{S}$. Typically, partial orders are used to compact the information of the orders given between the items in all the sequences. Note that posets suffice to summarize S: we do not require cycles in the structure since sequences in S correspond to total orders without self-loops. E.g., the poset in Figure 6.3(a) maps to each one of the sequences in $S = \{ \langle (A)(C)(G)(B)(E)(G)(F)(D) \rangle, \langle (E)(A)(C)(G)(F)(B)(D) \rangle \}$; or in other words, we may say that this partial order is compatible with S. Obviously, some partial orders summarize better a set S than others, and as it happened in the last chapter, from the universe of all the partial orders compatible with a set of sequences S, we are interested again in the most specific one. That is, we would like to construct a structure $\widehat{G}$ summarizing S and such that, for any other poset G compatible also with S, we have $G \trianglelefteq \widehat{G}$.

One way of constructing such a poset $\widehat{G}$ is by performing the n-ary cartesian product over the n sequences in S we want to summarize. Formally, we have a n-dimensional space associated to sequences $S = \{s_1, \ldots, s_n\}$, i.e. $\{(s_1[i_1], \ldots, s_n[i_n]) \mid i_1, \ldots, i_n\}$. From this space, we construct a partial order $\widehat{G}$ by associating a node to a set of positions $i_1, \ldots, i_n$ sharing exactly the same items (being this the label of the node), and by associating edges between nodes as long as they respect the forward direction.

Definition 6.7. Given a set of sequences $S = \{s_1, \ldots, s_n\}$, the **product of** S, denoted as $\prod S$, is a partial order $G_{\prod} = (V, G, \lambda)$ such that: $u_{i_1, \ldots, i_n} \in V$ if $s_1[i_1] = \ldots s_n[i_n]$, and the labelling function assigns $\lambda(u_{i_1, \ldots, i_n}) = s_1[i_1]$, and we have the edge $(u_{i_1, \ldots, i_n}, u'_{i'_1, \ldots, i'_n}) \in E$ when indexes are s.t. $i_1 < i'_1, \ldots,$ and $i_n < i'_n$.

We can prove that the product of the sequences in S is the terminal object from the universe of all the possible posets compatible with S. Formally:

Lemma 6.1. *For any partial order G compatible with S, we have that $G \trianglelefteq \prod S$.*

Proof. Let $h_s : G \mapsto s$ be a morphism from G to the total order obtained as the transitive closure of a sequence $s \in S$ (by hypothesis of compatibility, the corresponding morphism must exist for each $s \in S$). These morphisms h_s, $s \in S$, define for each node in G a set of positions over sequences in S as follows: for $v \in G$, we define a set of indexes $J = \{i \mid h_s(v) = s[i], s \in S\}$. Now, the morphism $g : G \mapsto \prod S$ will be s.t. $g(v) = u_J$, and u_J is one of the nodes in $\prod S$ by construction. Moreover, this morphism g is edge-preserving, that is: for each $(v, v') \in G$, it is the case that $g(v) = u_J$, $g(v') = u_{J'}$ and $J < J'$ (that is, $J = \{i_1, i_2, \ldots\}$ and $J' = \{i'_1, i'_2, \ldots\}$ and $i_1 < i'_1, i_2 < i'_2$ and so on), so that by construction we always have that $(u_I, u_{I'}) \in \prod S$. Then, the function g exists between any G and $\prod S$. □

A different way to understand this proof is to consider subsequences of each $s \in S$: again, let G be a partial order compatible with S, and let s' be a subsequence of s where G can be mapped exhaustively, that is, we choose $s' \subseteq s$ such that the morphism $G \trianglelefteq s'$ still exists, yet for no other $s'' \subset s'$ the morphism $G \trianglelefteq s''$ is defined. If we consider the set of these subsequences s', namely S', we have by construction that $G \trianglelefteq \prod S'$; on the other hand, $S' \preceq S$ so that $\prod S' \trianglelefteq \prod S$, and by composition we have $G \trianglelefteq \prod S$.

Although this poset $G_{\prod}$ is the most specific partial order summarizing S, it is not desirable to perform such construction in practice. The n dimensions defined by S may be large enough to make this approach completely unsuitable. For example, in a real dataset such as the one formed by customer shopping sequences, we would have that a subset of data S may be composed of thousands of transactions, thus defining not only an noninterpretative product but a highly intractable dimensional space. Moreover, an item occurring n_i times in each $s_i \in S$ would generate a $\prod_i n_i$ nodes in the final product poset.

6.2.1 Path-preserving Transformations

As in the last chapter, it would be good to characterize the set of maximal paths of this most specific partial order compatible with a set of sequences. However, the first problem we find is that identifying these maximal paths of the most specific partial order as the intersections of input sequences is not direct. An example to illustrate this problem is again the partial order of Figure 6.3(b). It is clear that if this poset was the most specific for a given set of sequences S, then the maximal path corresponding the sequence of labels $\langle (C)(G)(D) \rangle$ would always be subsumed in the intersection of sequences in S, by the longest sequence of labels $\langle (C)(E)(G)(D) \rangle$, which also corresponds to a maximal path of the same poset. Here, we start our analysis based on the following basic property.

Lemma 6.2. *Let them be given a set of sequences $S \subseteq \mathscr{S}$, and let $G = (V,E,\lambda)$ be any partial order compatible with the set S. Then, we have that: for all paths $t = u_1 \to u_2 \ldots \to u_n$ of G, there exists some $s' \in S'$ s.t. $\langle (\lambda(u_1))(\lambda(u_2))\ldots(\lambda(u_n)) \rangle \subseteq s'$, where $S' = \bigcap s, s \in S$.*

Proof. If $t = u_1 \to u_2 \ldots \to u_n$ is a maximal path of G, then there exists an injective morphism $t \trianglelefteq G$. Since G is defined to be compatible with the set of sequences S, we also have that $G \trianglelefteq s$ for all $s \in S$. By composition of the morphisms, we get $t \trianglelefteq s$, for all $s \in S$. This implies that $\langle (\lambda(u_1))(\lambda(u_2))\ldots(\lambda(u_n)) \rangle \subseteq s$, for all $s \in S$, and thus, $\langle (\lambda(u_1))(\lambda(u_2))\ldots(\lambda(u_n)) \rangle$ will be a subsequence of some maximal sequence contained in the intersection of those S, that is, $\langle (\lambda(u_1))(\lambda(u_2))\ldots(\lambda(u_n)) \rangle \subseteq s'$, for some $s' \in S'$.   □

This means that the intersection of sequences in S contains the necessary "information" to construct the compatible partial orders with S, and in particular, also the most specific of those candidates. Unfortunately, by considering exactly the intersections of sequences in S as maximal paths of a partial order G (as we did in chapter 5, where we had the restriction of nonrepeated labels) does not always suffice to outcome with the final most specific structure (e.g. the path corresponding to the sequence of labels $\langle (C)(G)(D) \rangle$ of the poset of Figure 6.3(b), that we mentioned above, can never correspond exactly with one intersection).

Based on Lemma 6.2, our goal can be stated as follows: let S' denote from now on the intersection of a nonempty set of sequences S, i.e. $S' = \bigcap s, s \in S$, and let $\widehat{G}$ be a partial order compatible with S with the property of being maximally specific, here we want to construct this $\widehat{G}$ out of S'. We will attempt at doing this with a path-preserving colimit transformation.

6.3 Characterization

We start our characterization by studying the nodes of this $\widehat{G}$ compatible with S. In the second part of our analysis we take care of joining conveniently the nodes with the suitable edges, by means of characterizing a colimit transformation.

A first observation following from Lemma 6.2 is that each node of this $\widehat{G}$ must correspond to a position of some of the sequences in S'. Indeed, different positions of different sequences in S' may correspond to the same node of the final structure $\widehat{G}$; these will represent the crossing points of the different paths in the poset. An approach to find such nodes is to check if the positions of sequences in S' can be safely identified.

The intuition is that each time we identify two nodes we are at a point of crossing paths, and from here, crossing-point nodes of $\widehat{G}$ can be deduced. Yet, there is the problem of finding which positions of sequences in S' correspond to a valid matching, and thus, will lead to the same node. Note that not all positions having the same label are good to be identified: e.g. for the set $S' = \{\langle(A)(C)(C)(C)(A)\rangle, \langle(C)(A)(C)(C)(A)\rangle\}$, the first (C) of $\langle(A)(C)(C)(C)(A)\rangle$ will not be matched with any (C) of the other sequence, otherwise we would get a partial order having more paths than the ones included in S' (and we would lose the compatibility with the sequences S, as stated by Lemma 6.2). So, when identifying positions of sequences in S', we need to take care to not add new paths to $\widehat{G}$ that are not represented in S'. The idea is that, given two sequences $s, s' \in S'$, two positions can be identified as long as they are path preserving with respect to all sequences in S'.

Definition 6.8 (Path Preserving Positions). Given is a set of sequences S'; let $s, s' \in S'$ be two sequences s.t. $s = \langle(I_1)\ldots(I_i)\ldots(I_n)\rangle$ and $s' = \langle(I'_1)\ldots(I'_j)\ldots(I'_m)\rangle$. Positions i of s and j of s' are path preserving if they satisfy the following three conditions:

- $I_i = I'_j$; and,
- $head(s, i) \diamond tail(s', j+1) \subseteq s''$, for some $s'' \in S$; and,
- $head(s', j) \diamond tail(s, i+1) \subseteq s''$, for some $s'' \in S$.

Then, we say that position i of s matches (or, is identified with) position j of s'; we denote this by $s[i] \sim s'[j]$.

The operation defined here is symmetric, i.e. if $s[i] \sim s'[j]$ then $s'[j] \sim s[i]$, but not necessarily transitive, that is, given $s, s', s'' \in S'$, if $s[i] \sim s'[j]$ and $s'[j] \sim s''[k]$, this does not necessarily imply that $s[i] \sim s''[k]$. By means of an example, let $S = \{\langle(A)(C)(G)(B)(E)(G)(F)(D)\rangle, \langle(E)(A)(C)(G)(F)(B)(D)\rangle\}$, whose intersection is $S' = \{\langle(A)(C)(G)(B)(D)\rangle, \langle(A)(C)(G)(F)(D)\rangle, \langle(E)(G)(F)(D)\rangle\}$. When looking only at the position where the item (G) is placed for the sequences in S', we see that: $\langle(A)(C)(G)(B)(D)\rangle \sim \langle(A)(C)(G)(F)(D)\rangle$ and also $\langle(A)(C)(G)(F)(D)\rangle \sim \langle(E)(G)(F)(D)\rangle$. However, $\langle(A)(C)(G)(B)(D)\rangle$ and $\langle(E)(G)(F)(D)\rangle$ do not match in the position where G is placed.

The final partial order that we would like to obtain for this example is the one depicted in Figure 6.3(a). For the moment, we will pay attention to the different nodes of the poset. The repeated node labelled with item G exactly corresponds to that group of positions in S' where the transitivity of path-preserving matchings does not hold. Formally, this can be characterized by describing each different label (we recall that for the moment each label is a single item, and later we will deal with simultaneous items) in our sequences with an undirected graph of positions, as follows.

Definition 6.9 (Position Graph). Given a set of sequences S', representing the intersection of S, and a label l, we define an undirected graph such that: for all $s \in S'$ and all positions i, we have that $s[i]$ is a node of the graph if $s'[i] = l$, and, there is an edge between the node $s[i]$ and $s'[j]$ if $s[i] \sim s'[j]$. The graph constructed in this way will be called the **position graph** for l in S'.

Fig. 6.4 Example of a position graph with three nodes

An example of a position graph with just three nodes is shown in Figure 6.4. In this case the graph is complete, meaning that the depicted positions for the sequences are path-preserving between them, that is, all the paths that will cross those positions are included in the set S'. The idea is that position graphs for a certain label will determine all the possible allowed combinations of path-preserving matchings for that label, and thus, a valid node of the final poset $\hat{G}$ we are trying to construct.

Moreover, whenever two different positions correspond to the same sequence, they cannot be path preserving. Indeed, we always have that $s[i] \sim s[j]$ is a forbidden edge of the position graph for two different $i \neq j$ of the same sequence $s \in S'$. Let us suppose that $i < j$, then by checking Definition 6.8 we see that $head(s, j) \diamond tail(s, i + 1)$ will be always a larger than the sequence s itself. But sequences in S' are all of them maximal by definition, since they were obtained by intersecting a set of sequences S; so, there is no other sequence in S' where $head(s, j) \diamond tail(s, i + 1)$ can be contained. Thus, $s[i]$ will never be connected to $s[j]$ in the position graph.

This simple representation of the position graph allows us to study the nodes of the final $\hat{G}$. First, the following property can be readily seen.

Proposition 6.1. *The maximal cliques of the position graph for label l in S' correspond to a maximal sets of positions of sequences in S' where transitivity holds.*

In other words, each maximal clique in this graph corresponds to a set of positions of different sequences $s' \in S'$ that are path-preserving between them, that is, all their crossing paths are included in the set S' by definition. So, all the positions corresponding to maximal cliques in the position graphs can be safely identified together into a single node of the poset because: first, they have the same label, and second, this will not add any extra path to the final structure.

Notice that when not having repeated items (as in chapter 5), the position graph of each label induces always a maximal clique. Indeed, we were in the case of posets that were always injectively labelled, so that different positions of sequences in S' with the same label always referred to the same node. There, the coproduct operation identified positions in sequences as long as their labels were exactly the same.

Definition 6.10. Given a set of sequences S' and a label l, we define the set l^w as the w-th maximal clique of the position graph obtained from label l over S'.

For convenience, l^w will contain the list of vertices in its maximal clique, that is, $l^w = \{s[i], s'[j], \ldots, s''[k]\}$, where $s, s', \ldots, s'' \in S'$. Note that S' is usually implied by the context, and it is not included in the notation to not overload it.

Following this description, our characterization will associate a new node u to each maximal clique of a position graph of a label l. When this is done for all possible labels we will obtain the desired set of nodes for $\widehat{G}$. To give more intuitions on this idea: let us consider again the partial order in Figure 6.3(a), and the three maximal sequences corresponding to its maximal paths, $S' = \{\langle (A)(C)(G)(B)(D) \rangle, \langle (A)(C)(G)(F)(D) \rangle, \langle (E)(G)(F)(D) \rangle\}$. The position graphs of labels A, C and F correspond each one to a complete undirected graph of two nodes, such as in Figure 6.5(a). The position graph of label D would correspond to a complete undirected graph of three nodes, such as the one depicted in Figure 6.5(b). Finally, the position graph of the conflictive label G corresponds to a undirected graph of three nodes with only two edges, as in Figure 6.5(c).

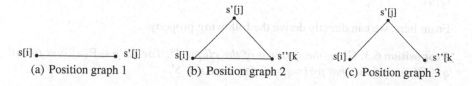

    (a) Position graph 1   (b) Position graph 2   (c) Position graph 3

Fig. 6.5 Example of three position graphs, that can be generated from the labels of the poset in Figure 6.3(a)

Then, nodes in the partial order of Figure 6.3(a) corresponding to labels A, C, F and D are unique since their corresponding position graphs are complete. However, the partial order has two nodes corresponding to label G, each one of them a maximal clique of its position graph.

6.3.1 Colimit Transformation

Let us consider now the set of those nodes stemming from each one of the maximal cliques of a label l for S', that is, for each one of the cliques l^w that we can obtain from the different labels l and indexes w, we associate a vertex. We want to construct a partial order structure out of them that corresponds exactly to the most specific poset for S. We will need the following definitions.

Definition 6.11. Let u be a node that is associated to the maximal clique $l^w = \{s[i], s'[j], \ldots, s''[k]\}$. We define the set prefixes and suffixes of the node as follows:

- $P(u) = \{head(s, i-1), head(s', j-1), \ldots, head(s'', k-1)\}$
- $Q(u) = \{tail(s, i+1), tail(s', j+1), \ldots, tail(s'', k+1)\}$

By construction, for all $p \in P(u)$ and for all $q \in Q(u)$, we always have that $p \diamond l \diamond q \subseteq s'$, for some $s' \in S'$.

That is, each $p \in P(u)$ corresponds to the prefix up to u of all sequences in S' participating in the same clique; and $q \in Q(u)$ to the suffixes. Note that we necessarily have that the sequences in $P(u)$ and $Q(u)$ are maximal, as this follows from the following property.

Proposition 6.2. *Let u be a node associated to l^w. Then, all sequences in $P(u)$ and $Q(u)$ are maximal sequences in the sets $P(u)$ and $Q(u)$ respectively.*

Proof. Each $p \in P(u)$ represents the head of a sequence $s' \in S'$ up to a certain position, say k; that is $p \diamond tail(s', k) \subseteq s'$. If we had $p \subset p'$ for some $p' \in P(u)$, by the construction given after Proposition 6.1 we would have also $p' \diamond tail(s', k) \subseteq s''$, for some $s'' \in S'$. Then, we obtain $s' \subset s''$ and both belong to S', which contradicts that S' contains only maximal sequences coming from the intersection of a nonempty set S. The same reasoning can be used to prove the maximality of sequences in $Q(u)$. $\square$

From here, we can directly derive the following property.

Proposition 6.3. *Given the vertex u of the clique l^w. For all $p \in P(u)$ and for all $q \in Q(u)$, we have that $p \diamond l \diamond q = s'$, for some $s' \in S'$.*

Proof. By definition of l^w, we have that $p \diamond l \diamond q \subseteq s'$, for some $s' \in S'$. Moreover, we must have that $p \diamond l \diamond q = s'$, otherwise either p or q would not be maximal in $P(u)$ or $Q(u)$ respectively. $\square$

The next step is to connect the different cliques, that is, vertices associated to the cliques l^w, with edges. This will be done through the colimit transformation. Before, we need to define the notion of valid edge between nodes.

Definition 6.12. An edge (u, u') between two nodes u and u', associated to the maximal cliques l^w and $l'^{w'}$ respectively, is said to be **path-preserving** if and only if for all $p \in P(u)$ and all $q \in Q(u')$ we have that $p \diamond l \diamond l' \diamond q \subseteq s'$, for some $s' \in S'$.

By construction, the following property will always hold.

Lemma 6.3. *Let u, u', u'' be three vertices associated to maximal cliques l^{w1}, l'^{w2} and l''^{w3} respectively. If edges (u, u') and (u', u'') are path-preserving, then the transitive edge (u, u'') is also path-preserving.*

Proof. Let us follow the notation $p' \in P(u')$, $q' \in Q(u')$ and $p'' \in P(u'')$, $q'' \in Q(u'')$. If (u', u'') is path-preserving, we have that $p' \diamond l' \diamond l'' \diamond q'' \subseteq s'$, for some $s' \in S'$. After Proposition 6.3 there must exist some q' s.t. $p' \diamond l' \diamond q' = s'$ (where s' is the sequence above). Then, we have that $p' \diamond l' \diamond l'' \diamond q'' \subseteq p' \diamond l' \diamond q'$, which implies that $l'' \diamond q'' \subseteq q'$.

From here, we have that for all $p \in P(u)$, the sequence $p \diamond l \diamond l'' \diamond q'' \subseteq p \diamond l \diamond q'$, which at the same time is a subsequence of $p \diamond l \diamond l' \diamond q'$. This last sequence is by hypothesis a subsequence of s' (because the edge (u, u') is path-preserving as stated by the proposition), and thus, we obtain that $p \diamond l \diamond l'' \diamond q'' \subseteq s'$, for some $s' \in S$, and so, (u, u'') is also a path-preserving edge.   □

The colimit transformation is meant to put all valid edges in place in the final partial order $\widehat{G}$. We will construct the colimit over all the possible path-preserving edges, by choosing conveniently the morphisms between the different structures participating in the diagram. We define the following transformation.

Definition 6.13 (Colimit Transformation). We define the colimit transformation as the colimit of the following family of graphs and morphisms (see Figure 6.6):

- The family of graphs $\{M_i\}$ containing one single node associated to a maximal clique, and no edge. That is, $M_i = (V, E, \lambda)$ s.t. $V = \{u\}$ and u comes from the maximal clique l^w (for a label l and index w), $E = \emptyset$ and $\lambda(u) = l$.
- The family of graphs $\{G_j\}$ containing one single path-preserving edge and the involved two nodes of that edge. That is, $G_j = (V, E, \lambda)$ s.t. $V = \{u, u'\}$ for u and u' associated to the maximal clique l^w and $l'^{w'}$ respectively, and $(u, u') \in E$ with (u, u') path-preserving, and $\lambda(u) = l$ and $\lambda(u') = l'$.
- The family of morphisms $\{m_{ij}\}$ where each $m_{ij} : M_i \mapsto G_j$ is defined as the mapping from node u in M_i to node u in G_j, if $u \in G_j$ and $u \in M_i$ and u comes from same maximal clique l^w in both G_j and M_i. There will not be any morphism m_{ij} for those G_j that do not contain any common node to M_i.

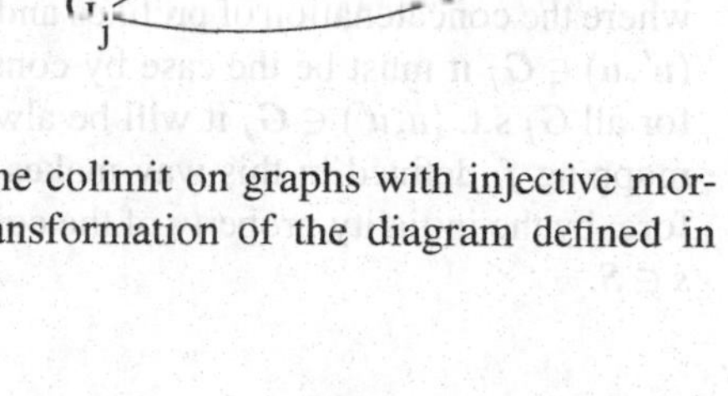

Fig. 6.6 Colimit transformation (from Definition 6.13)

It is well-known in category theory that the colimit on graphs with injective morphisms always exists. So, the colimit transformation of the diagram defined in Definition 6.13 exists (c.f. [1]).

6.3.2　Properties of the Colimit Transformation

Lemma 6.4. *Let $\widehat{G}$ be the partial order obtained from the colimit transformation of Definition 6.13; we have that $\widehat{G}$ is compatible with S.*

Proof. We will prove this statement based on the property that $\widehat{G}$ is, by definition, the initial object constructed by the colimit transformation. This means that any other poset G' (here, the transitive closure of each $s \in S$), that makes the diagram of the colimit commute, will be such that $\widehat{G} \trianglelefteq G'$ (see diagram of Figure 6.2). We show that for all $s \in S$, it is possible to define a morphism $f'_i : M_i \mapsto s$ and morphisms $h'_j : G_j \mapsto s$ that make the diagram of Figure 6.7 commute. Then, the initiality property of the colimit constructor will lead directly to $\widehat{G}$ having a morphism to s, for all $s \in S$.

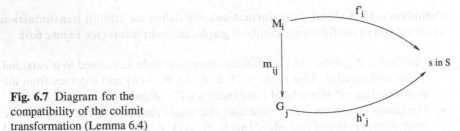

Fig. 6.7 Diagram for the compatibility of the colimit transformation (Lemma 6.4)

Let $u \in M_i$ with $\lambda(u) = l$ for a given M_i. After Proposition 6.3, we know that, for all $p \in P(u)$ and for all $q \in Q(u)$, the following is true: $p \diamond l \diamond q = s'$, for some $s' \in S'$. Because $s' \subseteq s$, and thus, a morphism from the total order s' to s exists, the result of Proposition 6.3 ensures that we can map each $p \in P(u)$ over s in such a way that the sequence s gets divided in two parts, namely $s = s_L \diamond l \diamond s_R$, such that $p \subseteq s_L$ and for all $q \in Q(u)$ we have that $q \subseteq s_R$. From all the prefixes in the set $P(u)$, we will choose that $p' \in P(u)$ generating the shortest right-hand side partition of s, that is, let $p' \in P(u)$ be the right-most subsequence over s so that: 1/ all the rest $p \in P(u)$, $p \subseteq s_L$ and still, 2/ all $q \in Q(u)$, are $q \subseteq s_R$.

Now, to make the diagram in Figure 6.7 commute, we simply need to define the mapping $f'_i : M_i \mapsto s$ that will divide the sequence s into the exact partition described above: we take $s = s_L \diamond l \diamond s_R$ s.t. for all $p \in P(u)$, $p \subseteq s_L$ and for all $p \in P(u)$, $q \subseteq s_R$. Here we define $f'_i(u)$ to map into the position of s identified by l. On the other hand, we know that edges of G_j are path-preserving by construction (Definition 6.12, where the concatenation of prefixes and suffixes are preserved) so that for all G_j s.t. $(u', u) \in G_j$ it must be the case by construction that $u' \trianglelefteq s_L$, and at the same time, for all G_j s.t. $(u, u') \in G_j$ it will be always the case that $u' \trianglelefteq s_R$. It is clear that the mapping f_i defined in this way makes the diagram in Figure 6.7 commute, therefore, by the initiality property of the colimit, there exists always a morphism $\widehat{G} \trianglelefteq s$, $s \in S$. □

The next final step is to prove that the partial order obtained from this colimit transformation is a most specific of such candidates compatible with S. A way to prove this result is to check that the universal structure defined from the product of sequences S, namely $\prod S$, has a morphism to $\widehat{G}$. If so, all the posets mapping to $\prod S$ will map to $\widehat{G}$ by composition of morphisms, and then, $\widehat{G}$ will be also a final object in the category of partial orders compatible with S.

This result is still unproved and we explicitly state it in Conjecture 6.1. The next section evaluates facts related to this proof and explains all the technical difficulties that make this conjecture remain still open. Also, we will show how to reduce this conjecture to a more technical question.

Conjecture 6.1. Let $\widehat{G}$ be the partial order obtained from the colimit transformation defined in Def. 6.13, we have that $\prod S \trianglelefteq \widehat{G}$.

6.3.3 Facts about Conjecture 6.1

Let $g : \widehat{G} \mapsto \prod S$ be an injective morphism from $\widehat{G}$ to $\prod S$. If this morphism exists, it will identify a set of nodes and edges over $\prod S$ corresponding to a reduced form of $\prod S$. Then, it will be possible to prove that any other poset G mapping to $\prod S$ (possibly having different ways of mapping), has at least one morphism $h : G \mapsto \prod S$, such that maps to this reduced subset of nodes and edges identified by g over $\prod S$. If this is the case, the morphism $G \trianglelefteq \widehat{G}$ of Conjecture 6.1 could be obtained by the composition of morphisms h and g^{-1}.

Indeed, we already have the following.

Lemma 6.5. *Let $\widehat{G}$ be the partial order obtained from the colimit transformation as defined above. There is always an injective morphism $\widehat{G} \trianglelefteq \prod S$.*

Proof. That the morphism $\widehat{G} \trianglelefteq \prod S$ exits is true by Lemma 6.1. Here we want to prove that there is always an injective mapping, named e.g. h, from $\widehat{G} = (\widehat{V}, \widehat{E}, \widehat{\lambda})$ to $\prod S = (V, E, \lambda)$. Let $u, v \in \widehat{G}$ s.t. $u \neq v$ and we assume to the contrary that $h(u) = h(v) = w$ for some $w \in \prod S$, i.e. nodes u and v are mapped to the same node w from the poset $\prod S$. This implies that edges connected to those nodes must be preserved also in $\prod S$, i.e. for all $(u_l', u) \in \widehat{G}$ and for all $(u, u_r') \in \widehat{G}$, we have that $(h(u_l'), w) \in \prod S$ and $(w, h(u_r')) \in \prod S$. Equally, this must be true for all the edges connected to v, i.e. for all $(v_l', v) \in \widehat{G}$ and for all $(v, v_r') \in \widehat{G}$, we have that $(h(v_l'), w) \in \prod S$ and $(w, h(u_r')) \in \prod S$. Yet, by the result given by Lemma 6.2 and the known compatibility of $\prod S$ with sequences in S, we know that any path crossing node w is indeed a subsequence of some $s' \in S'$. Thus, for all $p \in P(u)$ and for all $q \in Q(v)$, it is true that $p \diamond \widehat{\lambda}(u) \diamond \widehat{\lambda}(v) \diamond q$ is also included in some of the $s' \in S'$; in other words, the edge (u, v) is path-preserving, and it is also in $\widehat{G}$. Then, the morphism h is mapping two different nodes connected with an edge (u and v) to a single node of $\prod S$, and this mapping is not valid since self-loops do not exist in the structure by definition. So, we contradict the fact that two different nodes of $\widehat{G}$ can be mapped into one single node of $\prod S$, and therefore, the morphism between $\widehat{G} \trianglelefteq \prod S$ is necessarily injective. $\square$

Now, let $g : \widehat{G} \trianglelefteq \prod S$ be this injective morphism defined in Lemma 6.5. This morphism g divides the set of nodes of $\prod S$ in two categories. On the one hand, we have those nodes that belong to the codomain of the mapping function, i.e. nodes $u \in \prod S$ s.t. there exists a node $v \in \widehat{G}$ with $g(v) = u$. Broadly speaking, we can say that these nodes in $\prod S$ preserve, through g, the maximal clique identified in $\widehat{G}$. The other set of nodes from $\prod S$ are those not mapped by any node in $\widehat{G}$.

To prove our Conjecture 6.1, it will be important to keep track of those nodes in $\prod S$ that belong to the codomain of the mapping $\widehat{G} \trianglelefteq \prod S$. Our goal will be to find a "path-preserving" maximal clique in $\widehat{G}$ for each one of nodes of $\prod S$, in such a way that the whole structure of $\prod S$ is conveniently preserved by $\widehat{G}$ under homomorphism.

We propose the following definitions.

Definition 6.14. We say that a partition κ over a set of sequences $S = \{s_1, \ldots, s_n\}$ is a set of indexes $\kappa = \{k_1, \ldots, k_n\}$ within bounds, i.e. $1 \leq k_i \leq m_i$, where m_i is the length of sequence $s_i \in S$.

Here, each index k_i in κ will divide the sequence $s_i \in S$ in two parts: $s_i = head(s_i, k_i - 1) \diamond s_i[k_i] \diamond (s_i, k_i + 1)$. So, κ corresponds to a point in a S-dimensional space. As mentioned above, we will be interested in those nodes from $\prod S$, identified as maximal cliques of $\widehat{G}$ by morphism $\widehat{G} \trianglelefteq \prod S$, that are mapped always in the left-hand side of each $s \in S$, that is in the left-hand side of the partition defined by κ; reciprocally, we will define those nodes from $\prod S$, identified as maximal cliques of $\widehat{G}$ by morphism $\widehat{G} \trianglelefteq \prod S$, that are mapped in the right-hand side of the partition.

Definition 6.15. Given a partition $\kappa = \{k_1, \ldots, k_n\}$ over S, we define the following two sets of nodes:

- $\mathscr{C}_{L,\kappa} = \{u_J \in \prod S \mid u_J$ belongs to the codomain of $\widehat{G} \trianglelefteq \prod S$, and $J < \kappa\}$
- $\mathscr{C}_{R,\kappa} = \{u_J \in \prod S \mid u_J$ belongs to the codomain of $\widehat{G} \trianglelefteq \prod S$, and $\kappa < J\}$

To simplify notation, the capitalized index J of u_J refers to a set of indexes $J = \{j_1, \ldots, j_n\}$, so that $J < \kappa$ means $j_1 < k_1$, $j_2 < k_2$, $\ldots$, $j_n < k_n$.

Indeed, these last definitions can be geometrically justified as follows: each sequence $s \in S$ defines a component of the S-dimensional space of $\prod S$, and a partition κ corresponds to a point in this space. Then, $\mathscr{C}_{L,\kappa}$ defines the set of nodes in the left-hand side of the S-space that correspond to maximal cliques in $\widehat{G}$. Similarly, $\mathscr{C}_{R,\kappa}$ is the set of nodes in the right-hand side of the S-space corresponding to maximal cliques in $\widehat{G}$. A trivial property of partitions is the following.

Proposition 6.4. *Given two partitions κ_1 and κ_2 such that $\kappa_1 < \kappa_2$, we have that $\mathscr{C}_{L,\kappa_1} \subseteq \mathscr{C}_{L,\kappa_2}$ and $\mathscr{C}_{R,\kappa_2} \subseteq \mathscr{C}_{R,\kappa_1}$.*

Partitions in the S-space are meant to justify the final morphism required by Conjecture 6.1. As we formalized, any node $u_J \in \prod S$ has a set of indexes $J = \{j_1, \ldots, j_n\}$ indicating the exact mapping point of u_J over each one of the $s \in S$ (see Definition 6.7). This set of indexes J of each node in $\prod S$ can be seen as a partition of the S-space as well. As argued above, some of these nodes in $\prod S$ belong to the codomain of the injective morphism proved in Lemma 6.5, and this directly returns a partition in the S-space of the maximal cliques defining nodes in $\widehat{G}$.

The first useful result can be readily seen.

Proposition 6.5. *Let $\widehat{G}$ be the partial order obtained from the colimit transformation as defined above, and let $g : \widehat{G} \mapsto \prod S$ be the injective morphism proved in Lemma 6.5. For any given partition κ within the bounds of S, we have that for all $u \in \mathscr{C}_{L,\kappa}$ and $u' \in \mathscr{C}_{R,\kappa}$, there exists $(g^{-1}(u), g^{-1}(u')) \in \widehat{G}$.*

Proof. After the injective morphism g in Lemma 6.5 we have $g^{-1}(u) \in \widehat{G}$ and $g^{-1}(u') \in \widehat{G}$, for all $u \in \mathscr{C}_{L,\kappa}$ and $u' \in \mathscr{C}_{R,\kappa}$. To simplify, let us denote $g^{-1}(u)$ as the node $v \in \widehat{G}$ with label l, and $g^{-1}(u')$ as the node $v' \in \widehat{G}$ with label l'. Now, because both v and v' correspond to maximal cliques of $\widehat{G}$, it is true by Proposition 6.3 that: for all $p \in \mathsf{P}(v)$ and for all $q \in \mathsf{Q}(v)$, $p \diamond l \diamond q \subseteq s'$, for some $s' \in S'$; equally, for all $p' \in \mathsf{P}(v')$ and for all $q' \in \mathsf{Q}(v')$, $p' \diamond l \diamond q' \subseteq s'$, for some $s' \in S'$.

Also, we have by hypothesis that v is always mapped in front of v' in each one of the sequences $s \in S$ (because $g(v) \in \mathscr{C}_{L,\kappa}$ and $g(v') \in \mathscr{C}_{R,\kappa}$). Therefore, for all $p \in \mathsf{P}(v)$ and for all $q' \in \mathsf{Q}(v')$ we derive that $p \diamond l \diamond l' \diamond q'$ is a subsequence of some maximal sequence contained in the intersection of S, that is: $p \diamond l \diamond l' \diamond q' \subseteq s'$, for some $s' \in S'$. So, edge (v, v') is path-preserving and by construction of the colimit transformation $(v, v') \in \widehat{G}$.   □

This property can be generalized as follows.

Proposition 6.6. *Let $u_J \in \prod S$ and let the partition $\kappa = J$. We have that for all $v \in \mathscr{C}_{L,\kappa}$ and $v' \in \mathscr{C}_{R,\kappa}$, there exist $(v, u_J) \in \prod S$ and $(u_J, v') \in \prod S$.*

Proof. By definition each one of the nodes $v \in \mathscr{C}_{L,\kappa}$ belongs to $\prod S$ and by construction, v has an associated set of indexes v_I such that $I < \kappa$. Similarly, each one of the nodes $v' \in \mathscr{C}_{R,\kappa}$ has an associated set of indexes $v'_{I'}$ such that $\kappa < I'$. Therefore, by the structural construction of the product of S, we always have that $(v_I, u_J) \in \prod S$ and $(u_J, v'_{I'}) \in \prod S$, as we wanted to prove.   □

Finally, we can state the following important middle step towards Conjecture 6.1. This is a more technical way of rephrasing Conjecture 6.1, which remains open.

Conjecture 6.2. Let $u_J \in \prod S$ and let the partition be $\kappa = J$. Then, there is at least a node $w \in \widehat{G}$ such that for all $v \in \mathscr{C}_{L,\kappa}$ and for all $v' \in \mathscr{C}_{R,\kappa}$, there exists $(g^{-1}(v), w) \in \widehat{G}$ and $(w, g^{-1}(v')) \in \widehat{G}$, with $g : \widehat{G} \mapsto \prod S$ being the injective morphism proved in Lemma 6.5.

The result of Conjecture 6.2 states that given any node of $u_J \in \prod S$, there will be always at least a maximal clique defining a node $s \in \widehat{G}$ that will preserve the same input edges and same output edges of u_J. The trick of this proof lies in the combinatorial problem of justifying that a partition J, defined by $u_J \in \prod S$, will always have at least one $s' \in S'$ with associated maximal cliques in both sides of the partition J. If so, at least a sequence $s' \in S'$ will be crossing that part of the space defined by the partition, and there will exist a clique associated to a middle position of s' preserving the same edges as u_J (by Proposition 6.6).

Obviously, proving Conjecture 6.2 would imply that it is possible to find a way of mapping $\prod S$ on $\widehat{G}$, as we want for Conjecture 6.1. Eventually, even if Conjecture 6.1 is not formally justified, we will assume this as a working hypothesis and use the colimit construction as a good final heuristic to summarize a set of given input sequences. The next chapter shows that this works very well in practical cases.

Finally, notice that formally, in the general category of partial orders, there is no uniqueness of the most specific poset for S. Indeed, under our working hypothesis, we can consider both $\prod S$ and $\widehat{G}$ as equivalent posets, in the sense that a morphism from one to the other exists but the two structures are different. So, here we just have uniqueness under equivalence.

6.4 Lattice Isomorphy

The closure system of partial orders was already defined in the last chapter (section 5.3). The Galois properties proved in Proposition 5.2 hold also for the general definition of morphisms used here. Again, this implies the existence of the two closure operators: Ω and $\widehat{\Omega}$, which ensure the stability under intersection of our concepts (indeed, the intersection of closed posets could be formalized via products of category theory, as in chapter 5). An example is presented in Figure 6.9 for the data in Figure 6.8.

This closure system can be obtained by transforming the lattice of stable sequences. To prove this result it is necessary to assume that Conjecture 6.1 is true. This will be our working hypothesis.

Theorem 6.1. *Assuming Conjecture 6.1, the lattice of closed sets of sequences in $\mathscr{D}$ and the lattice of closed partial orders in $\mathscr{D}$ are isomorphic.*

Proof. By definition, $\phi(O) = S$ corresponds to the intersection of those input sequences in $\mathscr{D}$ identified by O. By assuming Conjecture 6.1, S can be transformed

Seq id	Sequence
d_1	$\langle (A)(C)(D)(A)(A)(C)(A) \rangle$
d_2	$\langle (C)(B)(C)(A)(C) \rangle$
d_3	$\langle (C)(B)(A)(B)(C)(C)(A)(A) \rangle$
d_4	$\langle (A)(B)(C)(D)(C)(C)(A)(A) \rangle$

Fig. 6.8 Example of a set of data $\mathscr{D}$ with repeated items

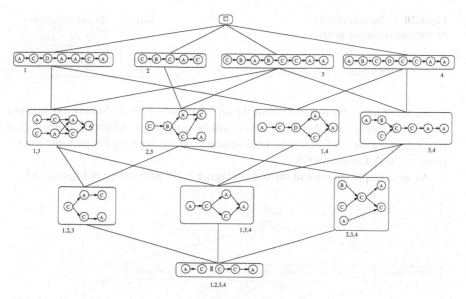

Fig. 6.9 Closure system of partial orders for the data in Figure 6.8

via colimit transformations on path-preserving edges into a partial order G s.t. any other G' compatible with input sequences identified by O satisfies $G' \trianglelefteq G$. Formally, this is sufficient to ensure that $\sigma(O) \trianglelefteq G$; also, we have that $G \trianglelefteq \sigma(O)$ by definition of σ. On the other hand, by definition of $\psi(S) = O$ the set of objects O is maximal for S, so that we necessarily have that $\tau(G) \subseteq O$; also, by definition of τ we know that $O \subseteq \tau(G)$. This implies $\tau(G) = O$. Then, (O, G) is a formal concept in the lattice of closed posets.

The other direction of the theorem is proved similarly. Now, let S be the set of sequences of labels corresponding to the maximal paths of the closed poset G. Some of the sequences in S might not be maximal now, so we consider $\max\{S\}$ (that is, only the sequences maximal in the set). Given that $\tau(G) = O$, we necessarily have that $\psi(S) = O$. Because S and $\max\{S\}$ are two sets of sequences occurring always in the same input transactions we also have, $\psi(\max\{S\}) = O$. Also, given that $\sigma(O) = G$, we necessarily have that $\phi(O) = \max\{S\}$. Thus, $(O, \max\{S\})$ is a formal concept in the lattice of stable sequences. □

6.4.1 The Simultaneity Condition Revisited

To consider input sequences $s = \langle (I_1)(I_2) \dots (I_n) \rangle$ where each I_i may contain several simultaneous items, we must redefine the labelling function l in the category of graphs with $\lambda : V \to 2^{\mathscr{I}}$. Indeed, each set of itemsets must be treated as an atomic label, and the construction of the position graph for a set of items will be made as usual over the label that they represent as a set. From this point of view, a label given

Fig. 6.10 Collection of data
$\mathscr{D}$ with simultaneous items

Seq id	Input sequences
d_1	$\langle (AE)(C)(D)(A) \rangle$
d_2	$\langle (D)(ABE)(F)(BCD) \rangle$
d_3	$\langle (D)(A)(B)(F) \rangle$

by the set of items I is different from the set I', whenever $I \neq I'$. So, even if we have two sets of items such that $I \subset I'$, they will be considered different atomic labels (so, no extra path is added to the final structure when checking for path-preserving positions with Definition 6.8).

As an example we present the data in Figure 6.10, with lattice in Figure 6.11.

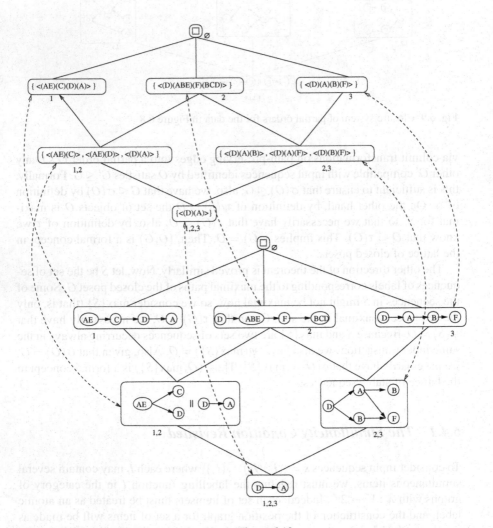

Fig. 6.11 Lattice transformation for data in Figure 6.10

An important remark to be done here regards the intersection of closed partial orders when simultaneity is considered. Following from Theorem 6.1 we have that the intersection of a set of closed posets can be defined in the isomorphic universe of closed sets of sequences (i.e. maximal paths of the poset corresponding to maximal sequences). From this point of view, the intersection of posets whose labels have simultaneous items should not consider each label as an atomic group of items anymore; indeed, labels should be allowed to be included in other labels (as is the case for the intersection of sequences). For example, the sequence $\langle (AE)(C) \rangle$ intersects with another sequence $\langle (D)(A)(F)(CD) \rangle$, and it returns $\langle (A)(C) \rangle$; if these sequences are seen as total orders, then sets of items in the labels must be treated as dividable.

Chapter 7
Towards Other Structured Data

Typically, research on pattern discovery has progressed from mining itemsets to mining sequences, and from sequences to mining other structures such as trees, lattices or graphs, e.g. a non-exhaustive set is [136, 6, 84, 13, 78, 83, 100, 134, 129, 140, 23, 24]. Last but not least, we raise the question of how to extend the results obtained with sequences to other structured data.

As we previously did with sequences and with partial orders, it would be still possible to define a proper Galois connection for graphs, or any other structured data. This would provide the closure operator from where to define a concept lattice on any context. The complexity of this formalization relies on the proper definition of the specificity relation and the intersection operation between the structured objects. For example, a first approach is the work of [92], where the authors define a formal framework to learning from examples described as labelled graphs, based on the use of a lattice. There, the necessary operations of specificity and intersection between labelled graphs are properly defined to prove the Galois properties, and thus, the contents of the nodes of the lattice correspond directly to the closed labelled graphs.

In practice, the direct construction of such lattices may be algorithmically expensive. This was also the case for constructing a lattice of closed partial orders on input sequences, yet there, we simplified the problem to closed sequential mining. In this chapter we propose a similar idea to generalize those results for structured objects. We explore the case of acyclic data, represented naturally as partial orders.

7.1 Acyclic Structured Data

We first consider the case of mining from data objects without cycles. This would be the case where the input transactions are trees, lattices, or general acyclic directed graphs. Indeed, any of these objects can be formalized as a partial order. As we saw in previous chapters, partial orders are a full subcategory of the set of directed graphs, and the formal advantage is that we already have available the corresponding set of morphisms to define specificity and generality between the structures. So,

G.C. Garriga: *Formal Methods for Mining Structured Objects*, SCI 475, pp. 85–97.
DOI: 10.1007/978-3-642-36681-9_7   © Springer-Verlag Berlin Heidelberg 2013

given two objects represented as two partial orders G and G', we say that G is more general than G' if there exists a morphism h from G to G', i.e. $h : G \mapsto G'$. These morphisms are not necessarily injective when dealing with the general category of graphs (then, we just have a preorder); injectivity can be considered only when not having repeated labels, as in chapter 5 (then, the morphism-based relationship defines a partial order).

An example of three partial orders is shown in Figure 7.1. We remind that edges belonging to the transitive closure of these objects are not depicted here, but formally they are contained in the definition. The new mining task consists of identifying those common substructures in the data over a certain user-specified threshold. The most specific poset common to all the structures of Figure 7.1 is the second poset in the same data, i.e. Figure 7.1(b).

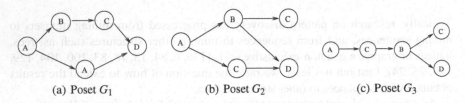

(a) Poset G_1 (b) Poset G_2 (c) Poset G_3

Fig. 7.1 An example of three input partial orders in the database

The algorithmic advantage of dealing with partial orders as our input data is that we do not need to mine directly these objects. Indeed, as suggested after results of chapter 5 and 6, each partial order can be represented by a set of sequences corresponding to its maximal paths. So, to deal with this problem, we propose to consider each partial order in our data as its set of maximal paths, i.e. each G will be transformed into a set of sequences $S = \{s_1, s_2, \dots, s_n\}$ where each s_i is a sequence of labels defining a maximal path of the partial order. E.g., the first object of data in Figure 7.1 corresponds to the set of sequences $S_1 = \{\langle (A)(B)(C)(D) \rangle, \langle (A)(A)(D) \rangle\}$.

After this transformation, the data to be mined is not anymore a set of partial orders, but it corresponds to a set of sets of sequences. Formally, the set of partial orders $\mathscr{D} = \{G_1, G_2, \dots, G_m\}$ of our database, is transformed into a new format data $\mathscr{D} = \{S_1, S_2, \dots, S_n\}$, where $S_i = \{s_{1i}, s_{2i}, \dots, s_{ni}\}$ is a set of sequences and each s_{ji} is a sequence of labels identifying a maximal path of the partial order G_i. An example for the transformation applied to the three objects in Figure 7.1 can be followed from the following table.

Of course, the subcategory of partial orders is wide enough to consider objects where paths are not uniquely identified by their sequence of labels. For example, in Figure 7.3(a) we modify slightly the poset of Figure 7.1(a). This modification yields a partial order where the sequence $\langle (A)(B)(C)(D) \rangle$ identifies two maximal paths of the structure. Also, we may consider the case where not all paths of the poset correspond to a maximal sequence of labels; e.g. the poset depicted in Figure 7.3(b), where the maximal path with sequence $\langle (A)(B)(D) \rangle$ is indeed included in

Fig. 7.2 Example of a transformed structured database $\mathscr{D}$

Poset Id	Sets of Maximal Paths
S_1	$\{\langle(A)(B)(C)(D)\rangle, \langle(A)(A)(D)\rangle\}$
S_2	$\{\langle(A)(B)(C)\rangle, \langle(A)(B)(D)\rangle, \langle(A)(C)(D)\rangle\}$
S_3	$\{\langle(A)(C)(B)(D)\rangle, \langle(A)(C)(B)(C)\rangle\}$

another sequence $\langle(A)(B)(C)(D)\rangle$, corresponding to another maximal path of the same partial order.

The correspondence of a partial order with its set of sequences of labels representing their paths is not unique; e.g. we may think of several possible posets stemming from the same set S. This will not be a problem for our later formalizations since we want to look for the most specific patterns in the data, so, just a nonredundant set of maximal sequences representing the paths suffice to profit from the previous theoretical results.

For coherence with the closure system of posets, we are assuming that all G in $\mathscr{D}$ is such that the colimit transformation on the sequences representing maximal paths of G returns a poset G' which is not necessarily equal to G but at least $G' \lhd G$. For the sake of illustration, the two orders from Figure 7.3 would correspond to the two sets: $S_a = \{\langle(A)(B)(C)(D)\rangle, \langle(A)(A)(D)\rangle\}$ for the partial order in 7.3(a), and the singleton $S_b = \{\langle(A)(B)(C)(D)\rangle\}$ for poset in 7.3(b). Then, the colimit of path-preserving edges on S_a returns a poset as the one depicted in Figure 7.1(a), which always has a morphism to the poset in Figure 7.3(a).

7.2 Mining a Set of Sets of Sequences

Now, the new problem corresponds to mining a set of sets of sequences, corresponding each one to a set of maximal sequences representing maximal paths. First, we formulate the new mining problem as follows:

Definition 7.1. A sequential pattern s occurs in $S \in \mathscr{D}$ when $\{s\} \preceq S$.

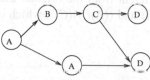

(a) A poset with two paths with exactly the same sequence of labels

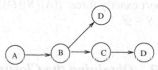

(b) A poset with a maximal path not being a maximal sequence of labels

Fig. 7.3 Another two examples of partial orders

That is, s occurs in the set of sequences $S \in \mathscr{D}$ if there exists at least one $s' \in S$ s.t. $s \subseteq s'$. Following data from Figure 7.2, we would say that a sequence such as e.g. $\langle (A)(D) \rangle$ occurs in S_1, S_2 and S_3. Then, we also say that the *support* of a sequence s is the number of $S \in \mathscr{D}$ where $\{s\} \preceq S$. The mining problem can be described as that of finding sequences s whose support is over a minimum threshold in the new transformed data $\mathscr{D}$.

Note that this new problem generalizes the sequential mining problem that we tackled in the previous chapters of this documentation. When mining sequences, we started from a database $\mathscr{D} = \{d_1, d_2, \dots d_n\}$, where each d_i corresponds to a single sequence; over this data, we were looking for sequential patterns s s.t. $s \subseteq d_i$, for $d_i \in \mathscr{D}$. When mining partial orders though, the data $\mathscr{D} = \{S_1, S_2, \dots, S_n\}$ corresponds to a set of sets of sequences, where each S_i represents the set of sequences of labels of a partial order G_i; in this new data, we are looking for sequential patterns s s.t. $\{s\} \preceq S$, for $S \in \mathscr{D}$.

Formally, we can regard this new problem as a generalization of the old one because each S in $\mathscr{D}$ may be also a set of one single sequence. This will be the case of having in $\mathscr{D}$ a set of input objects corresponding each one of them to a total order. Then, Definition 7.1 is equivalent to a simple subsequence comparison, as required in the sequential mining problem.

7.2.1 The New Closed Sequential Patterns

Correspondingly to any other mining problem, some of these frequent sequences in $\mathscr{D}$ may be redundant, so that we may restrict our search to the "closed" ones. Indeed, since the final obtained patterns will also be sequences, we can use the same definition of CloSpan (Definition 1.2), and that we renamed later as stable sequence. In other words, we will be interested in those patterns s such that no other supersequence s' has the same support. The semantic difference w.r.t. CloSpan is the notion of support, that now refers to support in sets of sequences, as explained above.

By way of an example, the sequence $\langle (A)(D) \rangle$ is not closed in data of Figure 7.2 since there exists the larger sequence $\langle (A)(B)(D) \rangle$ with exactly the same support in $\mathscr{D}$, that is, they both precede the same sets of sequences in the data. For this same example, there are three closed sequences, different from the original data, with support exactly three: $\langle (A)(B)(D) \rangle$, $\langle (A)(C)(D) \rangle$ and $\langle (A)(B)(C) \rangle$, which occur in each $S \in \mathscr{D}$.

7.2.2 Obtaining the Closure System

Interestingly, by reducing the partial order mining problem to sequences we enable all the results obtained in the previous chapters. First, we may realize that, as happens in the sequential case, there are closed patterns occurring simultaneously together in the same set of objects. For example, sequences $\langle (A)(B)(D) \rangle$,

$\langle (A)(C)(D) \rangle$ and $\langle (A)(B)(C) \rangle$ for data in Figure 7.2 represent a group of maximal closed sequential patterns occurring always together. Indeed, we can make use of the Algorithm 1 in chapter 3 to group the set of new frequent closed patterns into valid pairs. For this ongoing example, one of the valid pairs would correspond to the set $S = \{ \langle (A)(B)(D) \rangle, \langle (A)(C)(D) \rangle, \langle (A)(B)(C) \rangle \}$, that occurs in transactions $T = \{1,2,3\}$. Eventually, we will come up with a lattice of sets of sequences organized by the order $\preceq$, as we did in chapter 3.

By construction of the algorithm, we will have the following property.

Proposition 7.1. *Let* (T,S) *a valid pair constructed for the closed patterns of the transformed data* $\mathcal{D} = \{S_1, S_2, \ldots, S_n\}$ *with Algorithm 1 in chapter 3. Then:* $T = \{i | S \preceq S_i, S_i \in \mathcal{D}\}$, *and* $S = \{s | s$ *is maximal of those* $\{s\} \preceq S_i$, *for all* $i \in T\}$.

Moreover, we can naturally transform this lattice of valid pairs into a closure system as well, by intersecting the lists of object identifiers of formal concepts, as explained in the third chapter. Indeed, it is direct to realize that because this new problem generalizes the mining of sequences, we can generalize also the two derivation operators proposed in chapter 3, and from here, we can obtain the exact theoretical characterization of the general closure system.

Indeed, let $\mathcal{D} = \{S_1, S_2, \ldots, S_n\}$ be a set of sets of sequences. As customary in FCA, each S_i will have associated an object $o_i \in \mathcal{O}$ of the new context. When S_i contains one single sequence we will be working under ordered context, defined in chapter 3. Otherwise, we are working under a new structured context. We can generalize operators in chapter 3 as follows.

- For a set $O \subseteq \mathcal{O}$ of objects we generalize,
$$\widetilde{\phi}(O) = \{s \in \mathscr{S} | s \text{ maximal from those that } \{s\} \preceq S_o, \text{ for all } o \in O\}$$
- Correspondingly, for a set $S \subseteq \mathscr{S}$ of sequences we generalize,
$$\widetilde{\psi}(S) = \{o \in \mathcal{O} | S \preceq S_o\}$$

It is direct to realize that these two operators still satisfy the properties of the Galois connection. For the sake of illustration, the next proposition shows that they are dually adjoint (equivalent to proving the Galois properties [51]).

Proposition 7.2. $O \subseteq \widetilde{\psi}(S) \Leftrightarrow S \preceq \widetilde{\phi}(O)$.

Proof. Both directions can be simply proved as follows.

$\Rightarrow$/ Let $\widetilde{\psi}(S) = O'$. By construction of the operators, we have $S \preceq \widetilde{\phi}(O')$. Therefore, because $O \subset O'$ by hypothesis, we get: $S \preceq \widetilde{\phi}(O') \preceq \widetilde{\phi}(O)$.

$\Leftarrow$/ Let $\widetilde{\phi}(O) = S'$. By construction of the operators, we have $O \subseteq \widetilde{\psi}(S')$. Therefore, because $S \preceq S'$ by hypothesis, we get: $O \subseteq \widetilde{\psi}(S') \subseteq \widetilde{\psi}(S)$. $\qquad\square$

After this simple formalization we have set in place the theoretical characterization of the closure system on sets of sequences. We can proceed now to transform each node into a closed partial order.

7.2.3 Reconstructing Partial Orders

The most important consequence of the lattice construction and Proposition 7.1, is that the groups of patterns obtained from the new mining problem can be transformed into a higher level structure, by means of coproducts and colimits, shown in chapters 5 and 6. Formally, let (T, S) be a valid pair as explained above. By Proposition 7.1 we know that S corresponds to *all* the maximal paths common to *all* the objects identified by indexes in T. Then, the initiality of the colimit (or coproduct) construction on path-preserving edges obtained from S, returns always a partial order compatible with those input objects indexed by T. Relying on the results proved in the last chapters, this poset obtained will be the most specific one of all such candidates.

Fig. 7.4 Example of a transformation

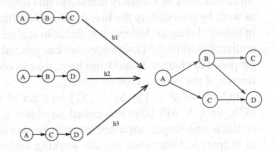

E.g., for the data in Figure 7.2 we mentioned that when considering support over two, we would get three frequent closed sequential patterns: $\langle(A)(B)(D)\rangle$, $\langle(A)(C)(D)\rangle$ and $\langle(A)(B)(C)\rangle$. The three of them occur always simultaneously in the same objects of $\mathscr{D}$, so they form a valid grouping, that is $\{\langle(A)(B)(D)\rangle, \langle(A)(C)(D)\rangle, \langle(A)(B)(C)\rangle\}$. Then, by considering that each one of these sequences represents indeed the maximal path of a higher level structure, we can transform it into the partial order of Figure 7.4, by using a coproduct here (because we do not have repeated items in this example). The object obtained in Figure 7.4 maps to all three objects in Figure 7.1.

If we consider the two partial orders of Figure 7.3, we see that the only closed sequence corresponds to $\langle(A)(B)(C)(D)\rangle$. From here we simply reconstruct a total order as $A \to B \to C \to D$. This total order has morphisms to the two original objects in Figure 7.3, and moreover, is the most specific one of all such candidates.

7.2.4 An Illustrative Example

Let us consider the three partial orders in Figure 7.5. These three posets could very well represent trees, for example, where the root of each tree would correspond to a source node. As proposed, first step is to transform each one of the input objects

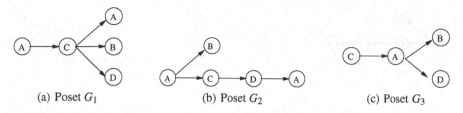

(a) Poset G_1 (b) Poset G_2 (c) Poset G_3

Fig. 7.5 An example of three input partial orders in our database

Fig. 7.6 Example of a transformed structured database $\mathscr{D}$

Poset Id	Sets of Maximal Paths
S_1	$\{\langle(A)(C)(B)\rangle, \langle(A)(C)(D)\rangle, \langle(A)(C)(A)\rangle\}$
S_2	$\{\langle(A)(B)\rangle, \langle(A)(C)(D)(A)\rangle\}$
S_3	$\{\langle(C)(A)(B)\rangle, \langle(C)(A)(D)\rangle\}$

T	S
$\{1,2,3\}$	$\{\langle(A)(B)\rangle, \langle(C)(D)\rangle, \langle(A)(D)\rangle, \langle(C)(A)\rangle\}$
$\{1,2\}$	$\{\langle(A)(B)\rangle, \langle(A)(C)(D)\rangle, \langle(A)(C)(A)\rangle\}$
$\{1,3\}$	$\{\langle(C)(B)\rangle, \langle(C)(A)\rangle, \langle(A)(B)\rangle, \langle(C)(D)\rangle, \langle(A)(D)\rangle\}$
$\{1\}$	$\{\langle(A)(C)(B)\rangle, \langle(A)(C)(D)\rangle, \langle(A)(C)(A)\rangle\}$
$\{2\}$	$\{\langle(A)(B)\rangle, \langle(A)(C)(D)(A)\rangle\}$
$\{3\}$	$\{\langle(C)(A)(B)\rangle, \langle(C)(A)(D)\rangle\}$

Fig. 7.7 Valid pairs of patterns considered for data in Figure 7.6, and using Definition 7.1

into a set of maximal sequences of labels representing its maximal paths. The transformation of these three partial orders can be followed from the table in Figure 7.6. Eventually, we will be mining the database $\mathscr{D} = \{S_1, S_2, \ldots, S_n\}$.

Once all these closed sequences are found, they can be grouped together into valid pairs, as it was done in chapter 3 with Algorithm 1. The set of all valid pairs is shown in the table of Figure 7.7.

From the list of valid pairs, we may also want to complete the closure system (as in chapter 3), by intersecting lists of objects and adding the missing concepts. The set of all concepts can be organized into a lattice structure of closed sets of sequences. Then, we can derive from each one of the concepts, the closed partial orders by using a colimit on path-preserving edges. This is depicted in Figure 7.8.

By the properties of the transformations, each one of the obtained partial orders has a morphism to the input posets identified by the associated list of transactions. Notice that the nodes located just below the top of the lattice generate exactly the same partial orders corresponding directly to our input data.

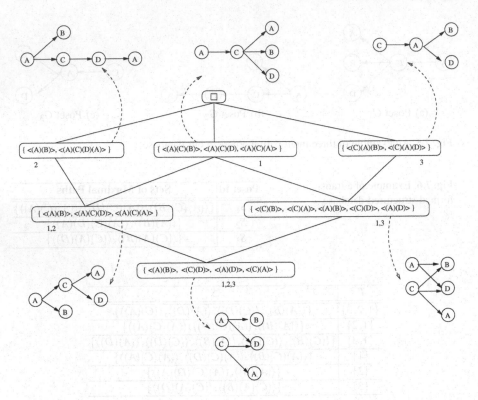

Fig. 7.8 Lattice of sets of sequences for data in Figure 7.6, and the derivation of partial order structures from each one of the nodes

7.3 An Integrated System for Structural Analysis

From the practical point of view a lattice of closed sets of sequences provides a natural framework to integrate different tasks for analyzing structured data. Indeed, all the results provided in this thesis are currently being implemented in a system named ISSA (Integrated System for Structural Analysis). This system is able to mine a set of partial orders as explained in this chapter. If the input posets are simply total orders we would be mining sequences, as in the previous chapters.

The preliminary implemented version of ISSA is programmed in C++ using STL libraries, under a Unix platform. For the moment ISSA has the restriction of not accepting sequences of itemsets, i.e. the simultaneity condition is not allowed on the input sequences. We consider that single-item sequences are a good start for the empirical validation since they model popular types of data, such as DNA, Web click streams, command histories of Unix users and so on. This preliminary version of ISSA is just a command line program with options allowing for the calculation of the different mining tasks on sets of sequences. Note that in most of the cases it is necessary to visualize the results, such as the final partial orders, or the hierarchical groupings of the input sequences. For those cases, ISSA outputs the necessary files in a standard GraphML format (see http://graphml.graphdrawing.org/), that can be visualized through any tool for graph visualizations, such as the Yed-Java Graph Editor (see http://www.yworks.com/), which is the one currently being used for our experimentations.

We distinguish three phases in the current system: 1/ identifying the frequent stable sequences from the lattice; 2/ organizing the frequent stable sequences into the lattice and outputting results for the desired mining task; and finally 3/ visualizing the results. So, in practice, we are constructing just the part of the lattice bounded by the stable sequences over a minimum support; if this threshold was set to zero, then, the full lattice would be depicted. The first phase is the most I/O intensive since it requires us to examine a computationally explosive number of patterns before identifying the stable ones. This can be done with current algorithms such as the mentioned CloSpan, BIDE or TSP.

The second phase of our approach requires the organization of the stable sequences into the lattice structure, as was explained in the third chapter. This is not an intensive step since the input to be examined is only the set of frequent stable sequences, and we do not require to combine those patterns, just to properly organize them into formal concepts. The operations performed to get such a lattice are simply standard operations of sets, such as inclusion, intersection and so on. The construction of the lattice and calculation of the necessary outputs usually takes a few seconds and a not significant use of memory. However, we noted that, when dealing with very large datasets, then the lists of sequence identifiers to be compared are long enough to make these operations more expensive.

Finally, the third phase is the visualization of the results such as the partial orders or the clustering of input sequences (that is, the lattice). Then, ISSA outputs the necessary GraphML file, so that we can visualize it by means of the Yed-Java Graph Editor, whose performance is not dependent of our program.

7.4 Experimental Validation

First, we evaluate the approach with a small database of 1382 transactions corresponding to the command history of a Unix computer user (downloaded from the UCI repository at http://kdd.ics.uci.edu/summary.data.type.html). Some statistics are shown in the following Table 7.1.

In this case we did not force the minimum support condition less than 20, since the number of discovered patterns clearly exceeded the 1382 initial transactions. As was to be expected, the number of partial orders coming from the nodes of the lattice, not forming a closure system, is always less than the number of stable sequences and frequent sequences. Some partial orders extracted from this dataset are shown in the piece of lattice of Figure 7.9. In the case that we are dealing with Unix

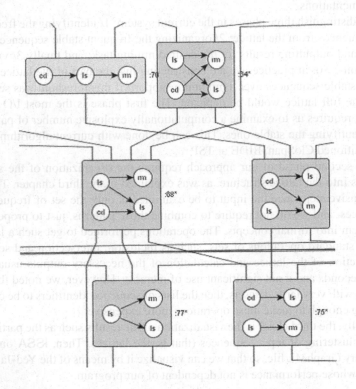

Table 7.1 Counting of patterns for the command Unix data

σ	Frequent Seqs.	Stable Seqs.	Closed Posets
50	176	175	175
30	782	765	762
20	3150	2892	2729

Fig. 7.9 Example of a small portion of the lattice with Unix command data

user data, the frequent closed partial orders may be later used as the normal user profile for the intrusion detection systems (such as proposed in [89]).

We also performed experiments with the msnbc.com anonymous web data provided again by the UCI repository. This data describes the page visits of users at msnbc.com: each sequence in the dataset corresponds to the page views of a user during a twenty-four hour period, and each event in the sequence corresponds to a user's request for a page. Requests are recorded by category, such as "frontpage", "news", "tech" ... We have a total of 989818 users recorded in the data.

In this lattice, we initially observe many recurrent partial orders with "frontpage" and "news" as a main events, that is, posets containing always the same event or the combinations of these two events. Broadly, the recurring posets reflect the revisiting of the front page and the news continuously during the twenty-four hour period of the same session: e.g. users would visit the news page to check the last updates, or

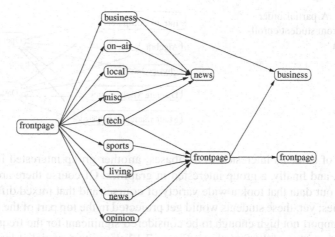

Fig. 7.10 A partial order derived from the msnbc.com anonymous web data

would unavoidably pass by the front page before clicking into a new category. Interestingly enough, one of the most significant partial orders that we find in the lattice is depicted in Figure 7.10. It is supported by 80500 users and it suggests that those users visit almost all the categories during their session: they always starting from the frontpage and then, there exists a clear split between going to see the "news" or returning to the "frontpage" again to click into a new page.

Finally, we wanted to semantically evaluate the approach with more complex structures. We performed experiments on a data set that describes the partial order in which about 12000 students of the Faculty of Informatics (FIB), at the Universitat Politècnica de Catalunya, registered to the courses during the last ten years. We run our three-step method for this data by imposing a minimum support threshold of 30%. Obtaining the final lattice with this support takes about five minutes; of course, with larger datasets the algorithmic efficiency would tend to degrade. At a glance, we observe important combinations of partial orders such as the one presented in Figure 7.11. These partial orders show that the introductory courses corresponding to the two first semesters exhibit a trivial order among them, but they occur in the data before the courses corresponding to the third and fourth semester. This behaviour can be explained by the filtering system that the university imposes between the first and second year: students are forced to succeed in all the introductory courses before stepping forward to the second-year courses. On the other hand, the trivial order exhibited among the courses of the first and second semester, is justified by the high rate of failure that we have during the first year. So, eventually students end up mixing courses of these two first semesters and breaking any pre-specified order.

By generating the concept lattice with the optional courses only (i.e. those chosen by the students, where we expect a lower failure rate), we observe that the lattice structure shows three clear tendencies separating students into three main groups:

Fig. 7.11 A partial order derived from student enrollment data

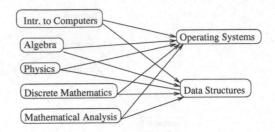

a group of students interested in databases, another group interested in operating systems, and finally, a group interested in graphics. Of course there are other students in our data that took a wide variety of courses and that mixed different other disciplines; yet, these students would get projected in the top part of the lattice, thus with a support not high enough to be considered significant for the frequent pattern mining algorithm of the first phase. Figure 7.12 shows two partial orders for two of these tendencies given by the lattice.

In general, the orders reflect the curriculum imposed by the university: e.g. students cannot register to "DB Management" until they do pass "DB Design". However, we also observe that some subjects, such as "Economy" or "Management of Projects", go beyond this preestablished order imposed by the curriculum. Both "Economy" and "Management of Projects" belong to the discipline of management and business, so are not related to databases or operating systems; however, the partial orders indicate that these are good complementary courses for any discipline. Another interesting observation stemming from the lattice corresponds to those students working for the master thesis (i.e. they are studying for a 5-year degree, in comparison to those other students that go for a 3-year degree), mainly belong to the group of students choosing databases as a discipline. This can be detected by including the master thesis as a subject in the database, and checking the partial orders containing it, e.g. as Figure 7.13.

Finally, we also decided to keep track of those nodes that make the system stable under intersection, that is, those extra nodes that the algorithm added to ensure a closure system. We find many practical cases where those nodes summarize perfectly a tendency that would not be visualized by the immediate predecessors in the lattice. For example, the partial order from Figure 7.13 stems from the union of

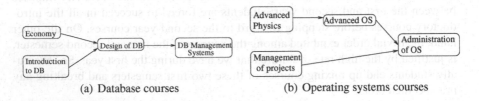

        (a) Database courses   (b) Operating systems courses

Fig. 7.12 Partial orders derived from the student enrollment data

Fig. 7.13 A partial order derived from student enrollment data

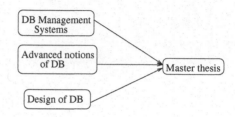

three immediate predecessors in the lattice containing a total order each. This part of the lattice is depicted in Figure 7.14. The unifying node from where we derive the final order, contains new information not included individually in the immediate predecessors.

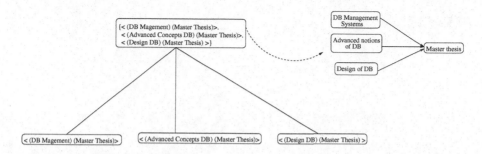

Fig. 7.14 A portion of the lattice for the student enrollment data

As a final remark, the lattice turns out to be a complete system to integrate different structural mining tasks. The main gain of our proposals relies on the possibility of generating classical partial orders out of the closed sequential patterns. The experiments on real datasets are semantically very appealing to understand the data. Finally, note that the main interest of our proposal is not to compare the efficiency of the method to other previous works, but to present a new intuitive way to generate partial order structures out of general data, which is totally based on a sound theory.

Chapter 8
Summary of Results

This manuscript proposed a formal framework for the mining of structured data. In the current state of the art, any algorithmic solutions have been proposed for this problem, yet little work has been done to formalize the underlying theory. We hope that the contributions presented here provide a better insight into the understanding of sequences, as well as other complex objects. The goal of this final chapter is to give a brief overview of the obtained results.

Galois Lattice of Closed Sets of Sequences. We introduce the basis of a framework for mining general patterns from a set of sequences. We develop on the hypothesis that the most representative of these patterns are the closed ones; so we characterize a new concept lattice for sequences in terms of a Galois connection. The new lattice model turns out to have quite useful properties. First, we have shown that the individual plain stable sequences in the concepts of the model exactly characterize the closed sequential patterns mined by CloSpan (and other related algorithms). We also analyze the construction of such a lattice in practice, and prove that just the proper organization of closed sequential patterns is sufficient to get the final closure system. Interestingly enough, this system turns out to be combinatorially very powerful; from here we develop the rest of our contributions related to sequences and partial orders.

The Full Implicational System for Sequences. We propose a notion of deterministic association rules in ordered data, building on the fact that such rules for the binary case can be formally justified as implications in a propositional logic framework. Our extension provides a way of mining facts where a set of subsequences implies another subsequence in the data, and proves that the final rules can be formally justified as well by a purely logical characterization, namely the empirical Horn approximation for ordered data. We reach this result by using the previously formalized closure system on sequences and the proper notion of minimal generators. We also discuss the algorithmic consequences of deriving such implications with order.

G.C. Garriga: *Formal Methods for Mining Structured Objects*, SCI 475, pp. 99–100.
DOI: 10.1007/978-3-642-36681-9_8   © Springer-Verlag Berlin Heidelberg 2013

Identification of Partial Orders in Sequences. We have presented the notion of closed partial orders compatible with a set of sequences of itemsets. By definition, these orders are the most informative ones for a set of maximal transactions and they provide a more compact overview of the input data. As a main contribution, we show that these closed partial orders can be derived from the closed sets of sequences represented in our lattice.

Initially, our analysis develops on the properties of the maximal paths that will compose the final closed partial orders. For the basic case of not having repeated items in the input sequences, we prove that these maximal paths of each closed poset correspond exactly to the intersection of those input sequences where the poset is compatible. For the general case this equivalence is no longer unique, so that the notion of path-preserving edges from those intersections is necessary to identify the most specific partial order. The final transformation of path-preserving edges into a higher-level structure resorts to two basic operations of category theory: coproducts when labels of the poset are not repeated, and colimits for the general model. In this latter case, we leave open the property that shows the maximal specificity of the final obtained poset. Finally, we prove a lattice isomorphy: the closure system of sequences can be transformed into a lattice of closed posets, and vice versa.

Towards Other Complex Structured Objects. The analysis of sequences and partial orders through FCA sets up the basis for the analysis of other complex structured data, represented here as partial orders. We propose a way to understand the mining of acyclic graphs by transforming each complex object into a set of sequences. Then, the use of a proper definition to match patterns in the new data suffices to rely on the previous theoretical results and construct hybrid structures. A proper empirical validation shows the viability of the proposal. This turns out to be an intuitive tool for analyzing real structured data.

References

1. Adamek, J., Herrlich, H., Strecker, G.: Abstract and Concrete Categories. John Wiley, New York (1990)
2. Agarwal, R.C., Aggarwal, C.C., Prasad, V.V.V.: A tree projection algorithm for generation of frequent item sets. J. Parallel Distrib. Comput. 61(3), 350–371 (2001)
3. Agrawal, R., Imielinski, T., Swami, A.N.: Mining association rules between sets of items in large databases. In: Buneman, P., Jajodia, S. (eds.) Proceedings of the 1993 ACM SIGMOD Int. Conference on Management of Data, pp. 207–216 (1993)
4. Agrawal, R., Mannila, H., Srikant, R., Toivonen, H., Verkamo, A.I.: Fast discovery of association rules. In: Advances in Knowledge Discovery and Data Mining, pp. 307–328 (1996)
5. Agrawal, R., Srikant, R.: Mining sequential patterns. In: Proceedings of the 11th Int. Conference on Data Engineering, pp. 3–14 (1995)
6. Arimura, H., Uno, T.: An Output-Polynomial Time Algorithm for Mining Frequent Closed Attribute Trees. In: Kramer, S., Pfahringer, B. (eds.) ILP 2005. LNCS (LNAI), vol. 3625, pp. 1–19. Springer, Heidelberg (2005)
7. Atallah, M.J., Gwadera, R., Szpankowski, W.: Detection of significant sets of episodes in event sequences. In: Proceedings of the 4th Int. Conference on Data Mining, pp. 3–10 (2004)
8. Baixeries, J., Balcázar, J.L.: Characterization and Armstrong Relations for Degenerate Multivalued Dependencies Using Formal Concept Analysis. In: Ganter, B., Godin, R. (eds.) ICFCA 2005. LNCS (LNAI), vol. 3403, pp. 162–175. Springer, Heidelberg (2005)
9. Baixeries, J., Garriga, G.C.: Sampling strategies for finding frequent sets. Revue des sciences et technologies de l'information (RTSI) 17(1), 159–170 (2003)
10. Baixeries, J., Garriga, G.C., Balcázar, J.L.: Best-first strategies for mining frequent sets. Journal d'Extraction des connaissances et apprentissage 1(4), 100–106 (2002)
11. Balcázar, J.L.: Closure-based confidence boost in association rules. Journal of Machine Learning Research - Proceedings Track 11, 74–80 (2010)
12. Balcázar, J.L.: Redundancy, Deduction Schemes, and Minimum-Size Bases for Association Rules. Logical Methods in Computer Science 6(2) (2010)
13. Balcázar, J.L., Bifet, A., Lozano, A.: Mining frequent closed rooted trees. Machine Learning 78(1-2), 1–33 (2010)
14. Balcázar, J.L., Baixeries, J.: Discrete deterministic datamining as knowledge compilation. In: SIAM Int. Workshop on Discrete Mathematics and Data Mining (2003)

15. Balcázar, J.L., Casas-Garriga, G.: On Horn Axiomatizations for Sequential Data. In: Eiter, T., Libkin, L. (eds.) ICDT 2005. LNCS, vol. 3363, pp. 215–229. Springer, Heidelberg (2005)
16. Balcázar, J.L., Garriga, G.C.: On Horn axiomatizations for sequential data. Theoretical Computer Science 371(3), 247–264 (2007)
17. Balcázar, J.L., Garriga, G.C., Díaz-López, P.: Reconstructing the rules of 1D cellular automata using closure systems. In: Proceedings of the 2nd European Conference on Complex Systems, pp. 55–61 (2005)
18. Bastide, Y., Pasquier, N., Taouil, R., Stumme, G., Lakhal, L.: Mining Minimal Non-redundant Association Rules Using Frequent Closed Itemsets. In: Palamidessi, C., Moniz Pereira, L., Lloyd, J.W., Dahl, V., Furbach, U., Kerber, M., Lau, K.-K., Sagiv, Y., Stuckey, P.J. (eds.) CL 2000. LNCS (LNAI), vol. 1861, pp. 972–986. Springer, Heidelberg (2000)
19. Bastide, Y., Taouil, R., Pasquier, N., Stumme, G., Lakhal, L.: Mining frequent patterns with counting inference. SIGKDD Explor. Newsl. 2(2), 66–75 (2000)
20. Bayardo, R.: Efficiently mining long patterns from databases. SIGMOD Rec. 27(2), 85–93 (1998)
21. Bayardo, R., Agrawal, R.: Mining the most interesting rules. In: ACM SIGKDD Int. Conference on Knowledge Discovery and Data Mining, pp. 145–154 (1999)
22. Berzal, F., Blanco, I., Sánchez, D., Vila, M.A.: Measuring the accuracy and interest of association rules: A new framework. Journal Intelligent Data Analysis 6, 221–235 (2002)
23. Bifet, A., Holmes, G., Pfahringer, B., Gavaldà, R.: Mining frequent closed graphs on evolving data streams. In: Proceedings of the 17th ACM SIGKDD Int. Conference on Knowledge Discovery and Data Mining (KDD 2011), pp. 591–599 (2011)
24. Borgelt, C., Meinl, T., Berthold, M.R.: Advanced pruning strategies to speed up mining closed molecular fragments. In: Proceedings of the IEEE Int. Conference on Systems, Man & Cybernetics (SMC 2004), pp. 4565–4570 (2004)
25. Boulicaut, J.F., Bykowski, A., Rigotti, C.: Free-sets: A condensed representation of boolean data for the approximation of frequency queries. Data Min. Knowl. Discov. 7(1), 5–22 (2003)
26. Brin, S., Motwani, R., Silverstein, C.: Beyond market baskets: Generalizing association rules to correlations. In: Proceedings of the ACM SIGMOD Int. Conference on the Management of Data, pp. 265–276 (1997)
27. Brin, S., Motwani, R., Ullman, J.D., Tsur, S.: Dynamic itemset counting and implication rules for market basket data. In: Proceedings ACM SIGMOD Int. Conference on Management of Data, pp. 255–264 (1997)
28. Burdick, D., Calimlim, M., Gehrke, J.: Mafia: A maximal frequent itemset algorithm for transactional databases. In: Proceedings of the 17th Int. Conference on Data Engineering, p. 443. IEEE Computer Society (2001)
29. Bykowski, A., Rigotti, C.: A condensed representation to find frequent patterns. In: Proceedings of the 20th ACM SIGMOD-SIGACT-SIGART Symposium on Principles of Database Systems (PODS 2001), pp. 267–273. ACM Press (2001)
30. Cadoli, M.: Knowledge compilation and approximation: Terminology, questions, references. In: 4th. Int. Symposium on Artificial Intelligence and Mathematics, AI/MATH 1996 (1996)
31. Calders, T., Goethals, B.: Mining All Non-derivable Frequent Itemsets. In: Elomaa, T., Mannila, H., Toivonen, H. (eds.) PKDD 2002. LNCS (LNAI), vol. 2431, pp. 74–85. Springer, Heidelberg (2002)

32. Calders, T., Goethals, B.: Minimal k-Free Representations of Frequent Sets. In: Lavrač, N., Gamberger, D., Todorovski, L., Blockeel, H. (eds.) PKDD 2003. LNCS (LNAI), vol. 2838, pp. 71–82. Springer, Heidelberg (2003)

33. Carpineto, C., Romano, G.: GALOIS: An order-theoretic approach to conceptual clustering. In: Proceedings of the 10th Int. Conference on Machine Learning, pp. 33–40 (1993)

34. Carpineto, C., Romano, G.: Effective reformulation of boolean queries with concept lattices. In: Proceedings of the Third Int. Conference on Flexible Query Answering Systems, pp. 83–94 (1998)

35. Carpineto, C., Romano, G.: Concept Data Analysis. Theory and Applications. Wiley (2004)

36. Carpineto, C., Romano, G., d'Adamo, P.: Inferring dependencies from relations: a conceptual clustering approach. Computational Intelligence 15(4), 415–441 (1999)

37. Chang, C., Keisler, J.: Model Theory. Elsevier, Amsterdam (1990)

38. Cong, S., Han, J., Padua, D.: Parallel mining of closed sequential patterns. In: Proceedings of the 11th ACM SIGKDD Int. Conference on Knowledge Discovery in Data Mining (KDD 2005), pp. 562–567. ACM Press (2005)

39. Cristofor, D., Cristofor, L., Simovici, D.A.: Galois Connections and Data Mining. Journal of Universal Computer Science 6(1), 60–73 (2000)

40. Cristofor, L., Simovici, D.: Generating an informative cover for association rules. In: Proceedings of the IEEE Int. Conference on Data Mining, pp. 597–600 (2002)

41. Das, G., Fleischer, R., Gasieniec, L., Gunopulos, D., Kärkkäinen, J.: Episode matching. In: Proceedings of the 8th Annual Symposium on Combinatorial Pattern Matching, pp. 12–27 (1997)

42. Davey, B.A., Priestly, H.A.: Introduction to Lattices and Order, Cambridge (2002)

43. Day, A.: The lattice theory of functional dependencies and normal decompositions. Int. Journal of Algebra and Computation 2(4), 409–431 (1992)

44. Demetrovics, J., Libkin, L., Muchnik, I.B.: Functional dependencies in relational databases: A lattice point of view. Discrete Applied Mathematics 40(2), 155–185 (1992)

45. Dong, G., Pei, J.: Sequence Data Mining (Advances in Database Systems). Springer-Verlag New York, Inc., Secaucus (2007)

46. Duquenne, V., Guigues, J.-L.: Famille minimale d'implication informatives résultant d'un tableau de données binaires. Mathématiques et Sciences Humaines 24(95), 5–18 (1986)

47. Eiter, T., Gottlob, G.: Identifying the minimal transversals of a hypergraph and related problems. SIAM J. Comput. 24(6), 1278–1304 (1995)

48. Fürnkranz, J., Flach, P.A.: An analysis of rule evaluation metrics. In: Proceedings of the 20th Int. Conference on Machine Learning, pp. 202–209 (2003)

49. Gallo, A., De Bie, T., Cristianini, N.: MINI: Mining Informative Non-redundant Itemsets. In: Kok, J.N., Koronacki, J., Lopez de Mantaras, R., Matwin, S., Mladenič, D., Skowron, A. (eds.) PKDD 2007. LNCS (LNAI), vol. 4702, pp. 438–445. Springer, Heidelberg (2007)

50. Ganascia, J.G.: TDIS: an algebraic formalization. In: Proceedings of the Int. Joint Conference on Artificial Intelligence, pp. 1008–1015 (1993)

51. Ganter, B., Wille, R.: Formal Concept Analysis. Mathematical Foundations. Springer (1998)

52. Garofalakis, M., Rastogi, R., Shim, K.: Spirit: Sequential pattern mining with regular expression constraints. In: Proceedings of the 25th Int. Conference on Very Large Data Bases, pp. 223–234 (1999)

53. Casas-Garriga, G.: Discovering Unbounded Episodes in Sequential Data. In: Lavrač, N., Gamberger, D., Todorovski, L., Blockeel, H. (eds.) PKDD 2003. LNCS (LNAI), vol. 2838, pp. 83–94. Springer, Heidelberg (2003)

54. Garriga, G.C.: Towards a formal framework for mining general patterns from structured data. In: 2nd Int. KDD Workshop on Multirelational Datamining, pp. 14–26 (2003)

55. Garriga, G.C.: Statistical strategies to remove all the uninteresting association rules. In: Proceedings of 16th European Conference on Artificial Intelligence, pp. 430–435 (2004)

56. Garriga, G.C.: Summarizing sequential data with closed partial orders. In: Proceedings of the SIAM Int. Conference on Data Mining, pp. 380–391 (2005)

57. Garriga, G.C.: Formal Methods for Mining Structured Objects. PhD Dissertation, Departament de Llenguatges i Sistemes Informàtics. Universitat Politècnica de Catalunya (2006)

58. Garriga, G.C., Balcázar, J.L.: Coproduct transformations on lattices of closed partial orders. In: Proceedings of 2nd Int. Conference on Graph Transformation, pp. 336–351 (2004)

59. Garriga, G.C., Díaz-López, P., Balcázar, J.L.: A lattice-based method for structural analysis. Research Report LSI-05-40-R, Universitat Politècnica de Catalunya (2005)

60. Garriga, G.C., Khardon, R., De Raedt, L.: On mining closed sets in multi-relational data. In: Proceedings of the 20th International Joint Conference on Artifical Intelligence, pp. 804–809 (2007)

61. Garriga, G.C., Kralj, P., Lavrac, N.: Closed Sets for Labeled Data. Journal of Machine Learning Research 9, 559–580 (2008)

62. Geerts, F., Goethals, B., Mielikäinen, T.: Tiling Databases. In: Suzuki, E., Arikawa, S. (eds.) DS 2004. LNCS (LNAI), vol. 3245, pp. 278–289. Springer, Heidelberg (2004)

63. Godin, R., Missaoui, R.: An incremental concept formation approach for learning from databases. Theoretical Computer Science 133, 387–419 (1994)

64. Goethals, B., Zaki, M.: Advances in frequent itemset mining implementations: report on fimi'03. SIGKDD Explor. Newsl. 6(1), 109–117 (2004)

65. Gonzalez, J., Holder, L.B., Cook, D.J.: Graph-based concept learning. In: FLAIRS Conference, pp. 377–381 (2001)

66. Gunopulos, D., Khardon, R., Mannila, H., Saluja, S., Toivonen, H., Sharma, R.S.: Discovering all most specific sentences. ACM Trans. Database Syst. 28(2) (2003)

67. Gunopulos, D., Mannila, H., Saluja, S.: Discovering all Most Specific Sentences by Randomized Algorithms. In: Afrati, F.N., Kolaitis, P.G. (eds.) ICDT 1997. LNCS, vol. 1186, pp. 215–229. Springer, Heidelberg (1996)

68. Guralnik, V., Karypis, G.: A scalable algorithm for clustering sequential data. In: Proceedings of the 1st Int. Conference on Data Mining, pp. 179–186 (2001)

69. Guralnik, V., Karypis, G.: Parallel tree-projection-based sequence mining algorithms. Parallel Comput. 30(4), 443–472 (2004)

70. Hamrouni, T., Yahia, S.B., Slimani, Y.: Prince: An Algorithm for Generating Rule Bases Without Closure Computations. In: Tjoa, A.M., Trujillo, J. (eds.) DaWaK 2005. LNCS, vol. 3589, pp. 346–355. Springer, Heidelberg (2005)

71. Han, J., Cheng, H., Xin, D., Yan, X.: Frequent pattern mining: current status and future directions. Data Mining and Knowledge Discovery 15(1), 55–86 (2007)

72. Han, J., Pei, J.: Mining frequent patterns by pattern-growth: methodology and implications. SIGKDD Explor. Newsl. 2(2), 14–20 (2000)

73. Han, J., Pei, J., Yan, X.: Sequential pattern mining by pattern-growth: Principles and extensions. Recent Advances in Data Mining and Granular Computing (Mathematical Aspects of Knowledge Discovery) (2005) (to appear)

74. Han, J., Pei, J., Yin, Y.: Mining frequent patterns without candidate generation. In: Proceedings of the ACM SIGMOD Intl. Conference on Management of Data, pp. 1–12. ACM Press (2000)
75. Harms, S., Deogun, J., Saquer, J., Tadesse, T.: Discovering representative episodal association rules from event sequences using frequent closed episode sets and event constraints. In: Proceedings of the 1st Int. Conference on Data Mining, pp. 603–606 (2001)
76. Hilderman, R.J., Hamilton, H.J.: Evaluation of Interestingness Measures for Ranking Discovered Knowledge. In: Cheung, D., Williams, G.J., Li, Q. (eds.) PAKDD 2001. LNCS (LNAI), vol. 2035, pp. 247–259. Springer, Heidelberg (2001)
77. Hipp, J., Güntzer, U., Nakhaeizadeh, G.: Algorithms for association rule mining: a general survey and comparison. SIGKDD Explor. Newsl. 2(1), 58–64 (2000)
78. Inokuchi, A., Washio, T., Motoda, H.: Complete mining of frequent patterns from graphs: Mining graph data. Mach. Learn. 50(3), 321–354 (2003)
79. Ji, X., Bailey, J., Dong, G.: Mining minimal distinguishing subsequence patterns with gap constraints. Knowl. Inf. Syst. 11, 259–286 (2007)
80. Kautz, H., Kearns, M., Selman, B.: Horn approximations of empirical data. Artificial Intelligence 74(1), 129–145 (1995)
81. Kavvadias, D., Papadimitriou, C.H., Sideri, M.: On horn envelopes and hypergraph transversals. In: Proceedings of the 4th International Symposium on Algorithms and Computation, ISAAC 1993, pp. 399–405 (1993)
82. Kryszkiewicz, M.: Closures of Downward Closed Representations of Frequent Patterns. In: Corchado, E., Wu, X., Oja, E., Herrero, Á., Baruque, B. (eds.) HAIS 2009. LNCS, vol. 5572, pp. 104–112. Springer, Heidelberg (2009)
83. Kuramochi, M., Karypis, G.: Finding frequent patterns in a large sparse graph. In: SIAM Int. Conference on Data Mining, SDM 2004 (2004)
84. Kuznetsov, S.O., Samokhin, M.V.: Learning Closed Sets of Labeled Graphs for Chemical Applications. In: Kramer, S., Pfahringer, B. (eds.) ILP 2005. LNCS (LNAI), vol. 3625, pp. 190–208. Springer, Heidelberg (2005)
85. Lane, T., Brodley, C.E.: Sequence matching and learning in anomaly detection for computer security. In: Proceedings AAAI 1997 Workshop on AI Approaches to Fraud Detection and Risk Management, pp. 43–49 (1997)
86. Laxman, S., Sastry, P.S., Unnikrishnan, K.P.: A fast algorithm for finding frequent episodes in event streams. In: Proceedings of the 13th ACM SIGKDD International Conference on Knowledge Discovery and Data Mining, KDD 2007, pp. 410–419 (2007)
87. Laxman, S., Sastry, P.S., Unnikrishnan, K.P.: Discovering frequent episodes and learning hidden markov models: A formal connection. IEEE Transactions on Knowledge and Data Engineering 17, 1505–1517 (2005)
88. Lee, W., Stolfo, S.J., Chan, P.K.: Learning patterns from unix process execution traces for intrusion detection. In: Proceedings AAAI 1997 Workshop on AI Approaches to Fraud Detection and Risk Management, pp. 50–56 (1997)
89. Lee, W., Stolfo, S.J., Mok, K.: A data mining framework for building intrusion detection models. In: Proceedings of the IEEE Symposium on Security and Privacy, pp. 120–132 (1999)
90. Lin, D.-I., Kedem, Z.M.: Pincer Search: A New Algorithm for Discovering the Maximum Frequent Set. In: Schek, H.-J., Saltor, F., Ramos, I., Alonso, G. (eds.) EDBT 1998. LNCS, vol. 1377, pp. 105–119. Springer, Heidelberg (1998)
91. Lin, J.L., Dunham, M.H.: Mining association rules: Anti-skew algorithms. In: Proceedings of the 14th Int. Conference on Data Engineering, Orlando, Florida, USA, pp. 486–493. IEEE Computer Society (1998)

92. Liquiere, M., Sallantin, J.: Structural machine learning with galois lattice and graphs. In: Proceedings of the 15th Int. Conference on Machine Learning, pp. 305–313 (1998)

93. Luxenburger, M.: Implications partielles dans un contexte. Math. Inf. Sci. Hum. 29(113), 35–55 (1991)

94. Mampaey, M., Tatti, N., Vreeken, J.: Tell me what I need to know: succinctly summarizing data with itemsets. In: Proceedings of the 17th ACM SIGKDD International Conference on Knowledge Discovery and Data Mining (KDD 2011), pp. 573–581 (2011)

95. Mannila, H., Meek, C.: Global partial orders from sequential data. In: Proceedings of the 6th Int. Conference on Knowledge Discovery in Databases, pp. 161–168 (2000)

96. Mannila, H., Toivonen, H., Verkamo, A.I.: Discovering frequent episodes in sequences. Data Mining and Knowledge Discovery 1(3), 259–289 (1997)

97. Martins-Antunes, C., Oliveira, A.L.: Sequential pattern mining algorithms: Trade-offs between speed and memory. In: PKDD 2004 Workshop on Mining Graphs, Trees and Sequences (2004)

98. Méger, N., Rigotti, C.: Constraint-Based Mining of Episode Rules and Optimal Window Sizes. In: Boulicaut, J.-F., Esposito, F., Giannotti, F., Pedreschi, D. (eds.) PKDD 2004. LNCS (LNAI), vol. 3202, pp. 313–324. Springer, Heidelberg (2004)

99. Nguifo, E.M., Njiwoua, P.: IGLUE: A lattice-based constructive induction system. In: Proceedings of the 9th Int. Conference on Tools with Artificial Intelligence, pp. 75–76 (1997)

100. Nijssen, S., Kok, J.N.: Efficient discovery of frequent unordered trees. In: First Int. Workshop on Mining Graphs, Trees and Sequences (2003)

101. Park, J.S., Chen, M.S., Yu, P.S.: Using a hash-based method with transaction trimming for mining association rules. IEEE Transactions on Knowledge and Data Engineering 9(5), 813–825 (1997)

102. Pasquier, N., Bastide, Y., Taouil, R., Lakhal, L.: Closed set based discovery of small covers for association rules. In: Proceedings of the 15th Int. Conference on Advanced Databases, pp. 361–381 (1999)

103. Pei, J., Han, J., Mao, R.: CLOSET: An efficient algorithm for mining frequent closed itemsets. In: ACM SIGMOD Workshop on Research Issues in Data Mining and Knowledge Discovery, pp. 21–30 (2000)

104. Pei, J., Han, J., Mortazavi-Asl, B., Pinto, H., Chen, Q., Dayal, U., Hsu, M.: PrefixSpan: mining sequential patterns by prefixprojected growth. In: Proceedings of the 17th Int. Conference on Data Engineering, pp. 215–224 (2001)

105. Pei, J., Liu, J., Wang, H., Wang, K., Yu, P.S., Wang, J.: Efficiently mining frequent closed partial orders. In: Proceedings of the Fifth IEEE International Conference on Data Mining, ICDM 2005, pp. 753–756 (2005)

106. Pei, J., Wang, H., Liu, J., Wang, K., Wang, J., Yu, P.S.: Discovering frequent closed partial orders from strings. IEEE Trans. on Knowl. and Data Eng. 18, 1467–1481 (2006)

107. Pfaltz, J.L.: Closure lattices. Discrete Mathematics 154, 217–236 (1996)

108. Pfaltz, J.L., Taylor, C.M.: Scientific knowledge discovery through iterative transformations of concept lattices. In: SIAM Int. Workshop on Discrete Mathematics and Data Mining, pp. 65–74 (2002)

109. Plantevit, M., Crémilleux, B.: Condensed Representation of Sequential Patterns According to Frequency-Based Measures. In: Adams, N.M., Robardet, C., Siebes, A., Boulicaut, J.-F. (eds.) IDA 2009. LNCS, vol. 5772, pp. 155–166. Springer, Heidelberg (2009)

110. De Raedt, L., Ramon, J.: Condensed representations for inductive logic programming. In: Proceedings of the 9th Int. Conference on Principles of Knowledge Representation and Reasoning (KR 2004), pp. 438–446 (2004)

111. Sahami, M.: Learning Classification Rules using Lattices. In: Lavrač, N., Wrobel, S. (eds.) ECML 1995. LNCS, vol. 912, pp. 334–346. Springer, Heidelberg (1995)
112. Savasere, A., Omiecinski, E., Navathe, S.B.: An efficient algorithm for mining association rules in large databases. In: Proceedings of 21th Int. Conference on Very Large Data Bases, pp. 432–444. Morgan Kaufmann (1995)
113. Selman, B., Kautz, H.: Knowledge compilation and theory approximation. Journal of the ACM 43(2), 193–224 (1996)
114. Shalizi, C.R., Haslinger, R., Rouquier, J.B., Klinkner, K.L., Moore, C.: Automatic filters for the detection of coherent structure in spatiotemporal systems
115. Silberschatz, A., Tuzhilin, A.: On subjective measures of interestingness in knowledge discovery. In: Proceedings of the 1st Int. Conference on Knowledge Discovery and Data Mining, pp. 275–281 (1995)
116. Smyth, P.: Clustering Sequences with Hidden Markov Models. In: Advances in Neural Information Processing Systems, vol. 9, pp. 648–654 (1997)
117. Srikant, R., Agrawal, R.: Mining Sequential Patterns: Generalizations and Performance Improvements. In: Apers, P.M.G., Bouzeghoub, M., Gardarin, G. (eds.) EDBT 1996. LNCS, vol. 1057, pp. 3–17. Springer, Heidelberg (1996)
118. Stumme, G.: Iceberg query lattices for datalog. In: Proceedings of the 12th Int. Conference on Conceptual Structures (ICCS 2004), pp. 109–125 (2004)
119. Tan, P., Kumar, V., Srivastava, J.: Selecting the right interestingness measure for association patterns. In: ACM SIGKDD Int. Conference on Knowledge Discovery and Data Mining, pp. 32–41 (2002)
120. Taouil, R., Bastide, Y., Pasquier, N., Lakhal, L.: Mining bases for association rules using closed sets. In: Proceedings of the 16th Int. Conference on Data Engineering (ICDE 2000), p. 307. IEEE Computer Society (2000)
121. Tatti, N.: Maximum entropy based significance of itemsets. Knowl. Inf. Syst. 17, 57–77 (2008)
122. Tatti, N., Cule, B.: Mining closed strict episodes. In: Proceedings of the 2010 IEEE International Conference on Data Mining, ICDM 2010, pp. 501–510 (2010)
123. Toivonen, H.: Sampling large databases for association rules. In: Proceedings of the 22th Int. Conference on Very Large Data Bases, pp. 134–145. Morgan Kaufmann Publishers Inc. (1996)
124. Tronicek, Z.: Episode matching. In: Combinatorial Pattern Matching, pp. 143–146 (2001)
125. Tzvetkov, P., Yan, X., Han, J.: TSP: Mining top-k closed sequential patterns. In: Proceedings of the 3rd IEEE Int. Conference on Data Mining, pp. 347–358 (2003)
126. Uno, T., Asai, T., Uchida, Y., Arimura, H.: An Efficient Algorithm for Enumerating Closed Patterns in Transaction Databases. In: Suzuki, E., Arikawa, S. (eds.) DS 2004. LNCS (LNAI), vol. 3245, pp. 16–31. Springer, Heidelberg (2004)
127. Vaillant, B., Lenca, P., Lallich, S.: A Clustering of Interestingness Measures. In: Suzuki, E., Arikawa, S. (eds.) DS 2004. LNCS (LNAI), vol. 3245, pp. 290–297. Springer, Heidelberg (2004)
128. Valtchev, P., Missaoui, R.: Building Concept (Galois) Lattices from Parts: Generalizing the Incremental Methods. In: Delugach, H.S., Stumme, G. (eds.) ICCS 2001. LNCS (LNAI), vol. 2120, pp. 290–303. Springer, Heidelberg (2001)
129. Vanetik, N., Gudes, E.: Mining frequent labeled and partially labeled graph patterns. In: Proceedings of the 20th Int. Conference on Data Engineering (ICDE 2004), p. 91. IEEE Computer Society (2004)
130. Vreeken, J., van Leeuwen, M., Siebes, A.: Krimp: mining itemsets that compress. Data Min. Knowl. Discov. 23(1), 169–214 (2011)

131. Wang, J., Han, J.: BIDE: Efficient mining of frequent closed sequences. In: Proceedings of the 19th Int. Conference on Data Engineering, pp. 79–90 (2004)

132. Wang, J., Han, J., Li, C.: Frequent closed sequence mining without candidate maintenance. IEEE Trans. Knowl. Data Eng. 19(8), 1042–1056 (2007)

133. Wang, J., Han, J., Lu, Y., Tzvetkov, P.: TFP: An efficient algorithm for mining top-k frequent closed itemsets. IEEE Trans. Knowl. Data Eng. 17(5), 652–664 (2005)

134. Yan, X., Han, J.: gSpan: Graph-based substructure pattern mining. In: Proceedings of the 2002 IEEE Int. Conference on Data Mining (ICDM 2002), p. 721. IEEE Computer Society (2002)

135. Yan, X., Han, J., Afshar, R.: CloSpan: Mining closed sequential patterns in large datasets. In: Proceedings of the Int. Conference SIAM Data Mining, pp. 166–177 (2003)

136. Yan, X., Zhou, X.J., Han, J.: Mining closed relational graphs with connectivity constraints. In: Proceedings of the Eleventh ACM SIGKDD International Conference on Knowledge Discovery in Data Mining, KDD 2005, pp. 324–333 (2005)

137. Zaki, M.: Generating non-redundant association rules. In: Proceedings of the 6th Int. Conference on Knowledge Discovery and Data Mining, pp. 34–43 (2000)

138. Zaki, M.: Scalable algorithms for association mining. IEEE Transactions on Knowledge and Data Engineering 12(3), 372–390 (2000)

139. Zaki, M.: SPADE: An efficient algorithm for mining frequent sequences. Machine Learning Journal, special issue on Unsupervised Learning 42(1/2), 31–60 (2001)

140. Zaki, M.: Efficiently mining frequent trees in a forest. In: Proceedings of the 8th ACM SIGKDD Int. Conference on Knowledge Discovery and Data Mining, pp. 71–80 (2002)

141. Zaki, M.: Mining non-redundant association rules. Data Mining and Knowledge Discovery: An Int. Journal 4(3), 223–248 (2004)

142. Zaki, M., Gouda, K.: Fast vertical mining using diffsets. In: Proceedings of the 9th ACM SIGKDD Int. Conference on Knowledge Discovery and Data Mining (KDD 2003), pp. 326–335. ACM Press (2003)

143. Zaki, M., Hsiao, C.: CHARM: An efficient algorithm for closed itemset mining. In: In 2nd. SIAM Int. Conference on Data Mining (2002)

144. Zaki, M., Ogihara, M.: Theoretical foundations of association rules. In: SIGMOD-DMKD Int. Workshop on Research Issues in Data Mining and Knowledge Discovery (1998)

145. Zaki, M., Parthasarathy, S., Ogihara, M., Li, W.: New algorithms for fast discovery of association rules. In: 3rd Intl. Conference on Knowledge Discovery and Data Mining, August 12-15, pp. 283–296. AAAI Press (1997)

146. Zhou, W., Liu, H., Cheng, H.: Mining Closed Episodes from Event Sequences Efficiently. In: Zaki, M.J., Yu, J.X., Ravindran, B., Pudi, V. (eds.) PAKDD 2010, Part I. LNCS, vol. 6118, pp. 310–318. Springer, Heidelberg (2010)

147. Zhu, F., Yan, X., Han, J., Yu, P.S., Cheng, H.: Mining Colossal Frequent Patterns by Core Pattern Fusion. In: Proceedings of the 2007 Int. Conference on Data Engineering (ICDE 2007), pp. 706–715 (2007)

Index